GENETICS AND BREEDING OF EDIBLE MUSHROOMS

GENETICS AND BREEDING OF EDIBLE MUSHROOMS

Edited by

SHU-TING CHANG
JOHN A. BUSWELL
Department of Biology
The Chinese University of Hong Kong
Shatin, New Territories

and

PHILIP G. MILES
Department of Biological Sciences
State University of New York at Buffalo
USA

Gordon and Breach Science Publishers
Switzerland Australia Belgium France Germany Great Britain
India Japan Malaysia Netherlands Russia Singapore USA

Published under license by Gordon and Breach Science Publishers S.A.

Gordon and Breach Science Publishers

Y-Parc
Chemin de la Sallaz
CH-1400 Yverdon, Switzerland

Post Office Box 90
Reading, Berkshire RG1 8JL
Great Britain

Private Bag 8
Camberwell, Victoria 3124
Australia

3-14-9, Okubo
Shinjuku-ku, Tokyo 169
Japan

58, rue Lhomond
75005 Paris
France

Emmaplein 5
1075 AW Amsterdam
Netherlands

Glinkastrasse 13-15
O-1086 Berlin
Germany

5301 Tacony Street, Drawer 330
Philadelphia, Pennsylvania 19137
United States of America

Library of Congress Cataloging-in-Publication Data

Genetics and breeding of edible mushrooms / edited by Shu-Ting Chang, John A. Buswell, and Philip G. Miles.
p. cm.
Papers presented at UNESCO regional workshop held at the Chinese University of Hong Kong, July 14-20, 1991.
Includes bibliographical references and index.
ISBN 2-88124-561-7 (hardcover)
1. Mushrooms, Edible--Breeding--Congresses. 2. Mushroom culture--Congresses. 3. Mushrooms, Edible--Collection and preservation--Congresses. I. Chang, S. T. (Shu-ting), 1930- II. Buswell, John A. III. Miles, Philip G.
SB352.87.G46 1993 92-16952
635.8--dc20 CIP

 Printed in the United States of America.

CONTENTS

PREFACE

There is a long history of mankind's use of mushrooms as a food source and for medicinal or tonic purposes, and there is evidence that the cultivation of mushrooms had its beginning around 600 A.D. The cultivation of mushrooms for human consumption currently amounts to approximately 3.8 million metric tons annually. This production has been achieved as a consequence of extensive basic research in mushroom biology and technological developments. Much present day research is directed towards obtaining a more complete understanding of the nutritional and medicinal benefits to be obtained from the consumption of mushrooms, whose cultivation is an outstanding example of bioconversion technology. Thus, the mushroom is of great interest to the food and drug industries. Since the spent compost can also be used as animal fodder, mushroom cultivation is also attracting the attention of the feed industry.

In order to perform the research that will increase our knowledge of mushrooms and be of value to the industries concerned, an essential first step is the collection and preservation of the mushroom cultures that are used for the breeding of high yielding and better quality strains. From July 14-20, 1991, a UNESCO regional workshop was held at the Chinese University of Hong Kong entitled Culture Collection and Breeding of Edible Mushrooms. The purpose of this workshop was to provide participants with information on the principles and techniques involved in the maintenance of culture collections and the breeding of edible mushrooms, and to familiarize them with the latest research methods and technological approaches used in these fields. The workshop also provided an occasion for younger scientists to discuss their research activities with more experienced scientists, and the opportunity for scientists from different countries to develop collaborative research. The workshop covered the following four topics:

(i) Preservation and Degeneration of Culture Collections
(ii) Genetic System and Breeding of Edible Mushrooms
(iii) Application of Protoplast Technology
(iv) Molecular Approaches to Breeding Programmes

This book is an outgrowth of the workshop, augmented by contributions from experts who did not attend this meeting. This imparts a more integrated and coherent approach and a greater appeal to a wider range of professionals.

For financial and moral support in the planning and implementation of the workshop, as well as this book, profound gratitude and sincere thanks are expressed to the United Nations Educational, Scientific and Cultural Organization (UNESCO), the Croucher Foundation and to the Chinese University of Hong Kong.

Finally, we would like to acknowledge the excellent support given by Mr. W.C. Chan in the technical processing of the manuscripts.

S.-T. Chang
J.A. Buswell
P.G. Miles

FOREWORD

As the growth of mushroom production continues to increase and spread throughout the world, it is a matter of considerable concern that the resources for scientific research, which secures the present and determines the future technologies, are being substantially reduced. This poses many challenges, but paramount is the need for mushroom scientists to collaborate and communicate openly and freely so that new knowledge can be exchanged, duplication of research can be avoided and the total world effort can be integrated. In this way the outputs from the diminished resources can be maximised.

This publication is an excellent example of communication, collaboration and integration among mushroom scientists on a subject area which is central to the continued development and progression of mushroom cultivation worldwide. As more information and knowledge emerges on the genetics and the factors which control the stability and productive potential of edible mushrooms, the methods of collecting and storing of cultures – whether it be for research or for the production of spawn – must take this new knowledge into account. Also the modern techniques of genetic engineering and molecular biology offer exciting possibilities in the breeding of new strains which hitherto has been difficult to achieve. These developments will influence greatly the future technologies which will be applied by cultivators.

The recent statistics on the production of edible mushrooms throughout the world highlight the rapid expansion of cultivation in Asian countries. With the realisation that edible mushrooms are useful foods and the process of cultivation is environmentally beneficial, a more diverse range of species is assuming commercial importance. Most of the methods adopted for the new cultivated species, such as *Pleurotus* spp., *Lentinus edodes, Flammulina velutipes, Ganoderma lucidum,* etc., are based on those initially developed and used in Asian countries. China, Japan and South Korea have been major producers for many years, but lately many other Asian countries are emerging as significant producers.

It is pleasing that many of the mushroom specialists in these emerging countries were present and fully participated in UNESCO's regional workshop. This kind of international participation by experts, specialists, educationalists, advisors and also a few cultivators and spawn producers would not have been possible without the support and assistance of UNESCO. In addition to UNESCO, the other food and development agencies of the United Nations, UNDP and FAO, have over many years contributed much to the initiation and development of mushroom culture in the emerging mushroom producing countries (notably in India, Bhutan, Nepal, Pakistan, Burma, Indonesia, Thailand, North Korea, Vietnam, Laos, Turkey, Trinidad and Tobago, Kenya and Mauritius).

The contributors, the participants, UNESCO and the Chinese University of Hong Kong are to be congratulated on organizing and conducting an invaluable workshop, and acknowledgement is due to Professor S.-T. Chang and Dr. J.A. Buswell for inspirational leadership, superb organization and making the arrangements for these papers to be published and made available to the world's mushroom scientists.

W.A. (Fred) Hayes
President
International Society for Mushroom Science

CHAPTER 1

MUSHROOM AND MUSHROOM BIOLOGY

Shu-Ting Chang

Department of Biology, The Chinese University of Hong Kong,
Shatin, New Territories, Hong Kong.

1. INTRODUCTION

Due to its pleasant flavour and substantial protein, the mushroom is a source of human food which can help to satisfy the basic human requirement for better nutrition. The medicinal properties of mushrooms also represent a relatively untapped resource for medical applications. Furthermore, the substrates used for mushroom cultivation are derived mainly from agricultural and industrial organic waste materials. It is because of these attributes that mushroom research and industries have been gaining more and more attention in recent years.

The use of mushrooms and mushroom cultivation have a long history in human development. Mushroom science, derived from the principles of microbiology, environmental engineering and fermentation technology (Chang & Miles, 1982), has developed in modern times to form the basis both for new cottage type industries and for highly developed industrial mushroom growing complexes. Biological efficiency, i.e., the yield of fresh mushrooms, in proportion to the weight of compost at spawning, can reach 100% in experimental tests, with 40-60% as a good average value per crop.

In overall view, the world production of cultivated edible mushrooms was 2,182 thousand tons and 3,763 thousand tons in 1986 and 1989/90,

TABLE 1. Comparison of 1986 and 1989/90 world production of cultivated edible mushrooms.

Unit: (metric ton x 1000)

Species	1986		1989/90		% increase
	Fresh wt.	%	Fresh wt.	%	
Agaricus bisporus / bitorquis	1,225	55.8	1,446	38.1	19.0
Lentinus edodes	320	14.7	402	10.6	25.6
Volvariella volvacea	178	8.2	207	5.5	16.3
Pleurotus spp.	169	7.8	909	24.0	437.9
Auricularia spp.	119	5.5	400	10.5	236.1
Flammulina velutipes	100	4.6	143	3.8	43.0
Tremella fuciformis	40	1.8	105	2.8	162.5
Pholiota nameko	25	1.1	53	1.4	112.0
Hericium erinaceus	-	-	90	2.4	
Hypsizigus marmoreus	-	-	22	0.6	
Grifola frondosus	-	-	7	0.2	
Others	10	0.5	10	0.3	
Total	2,176	100.0	3,794	100.2	74.4

Source: Chang & Miles (1991).

respectively (Table 1). In those 3 years, mushroom production increased by 72.5% or an annual increase of 24.5% (Chang & Miles, 1991). A comparison of production between 1986 and 1989/90 reveals that all cultivated mushroom species increased during that period, ranging from 16% for *Agaricus* up to 438% for *Pleurotus*. The second largest increase was 236% for *Auricularia*. However, the percentage of total world production of *Agaricus* and *Lentinus* mushrooms decreased as a consequence of the increase in production of the other cultivated edible mushroom species, in particular *Pleurotus* species. If 88.8 cents per pound, reported as the average price received by growers in U.S.A. in 1900-1991 (NASS, 1991), is used for purposes of estimating the value of the total world mushroom crop, the figure for the 1989/90 financial year totalled US$7,485,058,500.

It should be noted that an upward tendency in world production of cultivated edible mushrooms is clearly indicated in Fig. 1. There is a particularly sharp increase in growth over the last five years. This tendency is expected to continue in the future due to advances both in our basic knowledge of mushroom biology and in the practical technology associated with mushroom cultivation. The significant impact of mushroom research and production can be considered globally but must be implemented according to local materials, labour and climatic conditions.

2. WHAT IS A MUSHROOM ?

This is not a new question or a new issue. The word "**mushroom**" may mean different things to different people in various countries. Even in the literature, the term mushroom may also have different meanings (Table 2). It is my viewpoint that specialized studies, and the economic value of mushrooms have reached the point where an up-to-date definition of the term mushroom is now warranted. This will serve a useful purpose at a time when the number of cultivated mushroom species is increasing, when production of established cultivated mushrooms has also shown a steady increase (Fig. 1), and when an

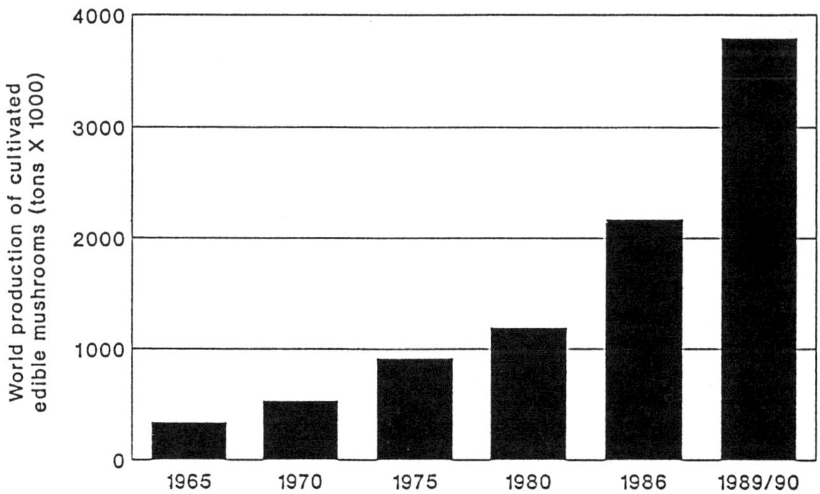

FIGURE 1. Annual world production of cultivated edible mushrooms.

increasing number of countries and people are engaged in mushroom cultivation as an agricultural or industrial technology. In this chapter, "**mushroom**" refers to the definition given by Chang and Miles (in press). In a broader sense "the mushroom is a macrofungus with a distinctive fruiting body which can be either epigeous or hypogeous and large enough

TABLE 2. The changing definition of mushroom.

Date	Source	Definition
1961	Atkinson, G.F.	The mushroom belongs to the basidiomycetes, and toadstool is regarded as a synonymous term, since there is, strictly speaking, no distinction between a mushroom and a toadstool.
1967	Gray, P.	An edible basidiomycete or, rarely, ascomycete fungus.
1971	Snell, W.H. & E.A. Dick	A mushroom may be edible, poisonous, unpalatable, tough, etc., but popular usage applies the term to edible ones, calling the others "toadstools".
1978	Chang S.T. & W.A. Hayes	Mushrooms refer to both epigeous and hypogeous fruiting bodies of macroscopic fungi.
1983	Pegler, D.N.	Mushroom and toadstools are terms rather loosely applied to the fruiting bodies of fleshy gill-fungi, and are commonly used to denote edible and poisonous species respectively.
1988	Mish, F.C.	An enlarged complex aerial fleshy fruiting body of a fungus (as of the class Basidiomycetes) that consisted typically of a stem bearing a flattened cap.
In press	Chang S.T. & P.G. Miles	The mushroom is a macrofungus with a distinctive fruiting body which can be either epigeous or hypogeous and large enough to be seen with the naked eye and to be picked by hand.

to be seen with the naked eye and to be picked by hand. Thus, mushrooms need not be Basidiomycetes, nor aerial, nor fleshy, nor edible. Mushrooms can be Ascomycetes, grow underground, have a non-fleshy texture and need not be edible". In other words, mushrooms can be roughly divided into four categories: (1) those which are fleshy and edible fall into the edible mushroom category, e.g., *Agaricus bisporus*; (2) mushrooms which are considered to have medicinal applications, are referred to as medicinal mushrooms, e.g., *Ganoderma lucidum*; (3) those which are proven to be, or suspected of being poisonous are named as poisonous mushrooms, e.g., *Amanita phalloides*; (4) a miscellaneous category which includes a large number of mushrooms whose properties remain less well defined. These may tentatively be grouped together as 'other mushrooms'. Certainly, this form of classifying mushrooms is not absolute. Many kinds of mushrooms are not only edible, but also possess tonic and medicinal qualities. In a botanical sense, mushrooms are fungi which lack chlorophyll and so cannot use solar energy to manufacture their own food as do green plants. However, mushrooms can produce a wide range of enzymes, that degrade the complex substrates on which they grow, following which they absorb the soluble substances for their own nutrition. This absorptive nutrition is a characteristic of fungi. Mushrooms can also be poetically described as: "Without leaves, without buds, without flowers: yet they form fruit. As a food, as a tonic, as a medicine: the entire creation is precious" (Chang 1990).

3. JUSTIFICATION FOR THE TERM MUSHROOM BIOLOGY

In any discipline, when knowledge increases and areas of specialization develop within the discipline, it is convenient to indicate that area of specialization with a self explanatory name. In biology, there are such specializations as neurobiology, bacteriology, plant pathology, pomology, molecular biology, virology, embryology, endocrinology, phycology, entomology/insect biology, plant biology and animal biology. These names indicate either a group of organisms (e.g., bacteria, algae, insects) and/or an approach to the study (e.g., disease, development, physiology).

Mycology is the science that deals with fungi, of which there are over 69,000 described species (Hawksworth, 1991). The fungi are of importance to man for a variety of reasons. They are the principal causal agents of plant diseases as well as some significant diseases of man. Through their fermentative activities, the fungi are major producers of some important products such as ethyl alcohol, citric acid, and the antibiotic, penicillin. Certainly not to be ignored is the edible mushroom, for the value of its annual production is estimated at about US$7.5 billion. Several terms for this important branch of mycology that deals with mushrooms have been used, and each of these has its merit. However, when we get down to the matter of definitions, it seems that there is a place for a new term, which is **Mushroom Biology**. Mushroom biology is the discipline concerned with the scientific study of mushrooms (Chang & Miles, in press). The term **mushroom science** already exists, but it is restrictive in that it has been defined as the discipline that is concerned with the principles and practices of mushroom cultivation. Mushroom biology includes not only cultivation but deals with every aspect of mushrooms, such as: taxonomy, development, nutrition, physiology, genetics, pathology, medicinal and tonic attributes, edibility, toxicity, etc. The activities centred around mushroom studies have recently achieved global dimensions and present many long term worldwide implications. These include conservation of mushroom germplasm as a part of the conservation of the world's biological diversity, which has emerged as a very serious matter of international concern (Miles & Chang, 1986; Deak, 1991).

TABLE 3. Comparison of the numbers of known and estimated total species in the world of selected groups of organisms.

Known Group	Known species	Total species	Percentage known (%)
Vascular Plants	220000	270000	81
Bryophytes	17000	25000	68
Algae	40000	60000	67
Fungi	69000	1500000	5
Bacteria	3000	30000	10
Viruses	5000	130000	4

Source: Hawksworth (1991)

TABLE 4. Genera of prime edible mushrooms.

Basidiomycetes	
Agaricus	*Lactarius*
Amanita	*Lentinus*
Armillaria	*Lepista*
Auricularia	*Lyophyllum*
Boletus	*Marasmius*
Cantharellus	*Pleurotus*
Calvatia	*Pholiota*
Clitocybe	*Polyporus (Grifola)*
Coprinus	*Russula*
Cortinarius	*Stropharia*
Dictyophora	*Termitomyces*
Flammulina	*Tremella*
Gloeostereum	*Tricholoma*
Hericium	*Volvariella*
Ascomycetes	
Morchella	*Tuber*

TABLE 5. Species of commercially cultivated edible mushrooms.

*Agaricus bisporus**	*Flammulina velutipes**
*Agaricus bitorquis**	*Lentinus edodes**
Auricularia auricula	*Lyophyllum ulmarium*
Auricularia polytricha	*Pholiota nameko*
Auricularia fuscosuccinea	*Pleurotus ostreatus**
Dictyophora indusiata	*Pleurotus sajor-caju*
Dictyophora duplicata	*Pleurotus cystidiosus*
Gloeostereum incarnatum	*Pleurotus cornucopiae*
Grifola frondosa	*Pleurotus florida*
Hericium erinaceus	*Stropharia rugoso-annulata*
Hypsizygus marmoreus	*Tremella fuciformis*
(=Pleurotus elongatipes and = *Lyophyllum shimeji)*	*Volvariella volvacea**

*Species produced on an industrial scale.

4. MAGNITUDE OF MUSHROOM SPECIES

The number of known species of fungi is about 69,000. It is conservatively estimated that 1.5 million species actually exist (Table 3) (Hawksworth, 1991). The fungi are regarded as being the second largest group of organisms in the biosphere after the arthropods, of which insects comprise by far the greatest number of species. The estimates of insects by Stork (1988) range between 10-80 million and by Thomas (1990) to 609 million. Known fungal species constitute only about 5% of the estimated total species in the world. Thus, the large majority of fungi are still unknown. Out of 69,000 described species of fungi, there are about 10,000 species of fleshy macrofungi and only a handful of these are lethal (Kendrick, 1985). There are no simple ways of distinguishing between edible and the poisonous mushrooms.

Mushrooms should be eaten only if they have been identified with precision. About 2,000 species from more than 30 genera (Table 4) are regarded as prime edible mushrooms, but only about 80 of them are grown experimentally, 40 cultivated economically, around 22 cultivated commercially (Table 5) and only 5 to 6 are produced on an industrial scale. In general, the oriental countries, China, Japan and Korea, grow and consume more varieties of mushrooms than the western countries. However, in recent years, the production of what are referred to as "specialty mushrooms", mainly *Lentinus edodes*, and *Pleurotus* spp., have increased rapidly in western countries (Chang & Miles, 1991). In the current context, "specialty" or "alternative" mushrooms are defined as cultivated mushrooms other than *Agaricus bisporus / bitorquis*.

5. CONCEPT OF MUSHROOM TECHNOLOGY

The consumption of mushrooms by man probably predates recorded history, and the historical record is, indeed, an ancient one. The historical records of the intentional cultivation of several important edible mushrooms are shown in Table 6 (Chang & Miles, 1987). It is estimated that the first mushroom was cultivated around 600 A.D. This was *Auricularia auricula*. Later, around 800-900 A.D., *Flammulina velutipes* was also cultivated in China. *L. edodes* is estimated by us to have been cultivated for the first time between 1000-1100 A.D. The mushroom produced in greatest amounts today, *A. bisporus*, was not cultivated until 1600. Of the leading mushrooms of today that were cultivated before 1900, *Agaricus* is the only one that was not first cultivated in China. Note that *Volvariella volvacea* is estimated to have been first cultivated around 1700 and *Tremella fuciformis* around 1800 - in China. Misconceptions about mushroom culture are exceedingly common, especially in the developing countries. Although it is thought to be very simple, mushroom cultivation is, in fact, a complicated business. It involves a number of different operations including preparation of a fruiting culture, spawn, and compost as well as crop management and marketing. While it can be treated as a primitive type of farming as in the cultivation of the straw mushroom, *V. volvacea*, in the Southeast Asia countries (Vedder, 1978; Chang, 1980;

TABLE 6. Historical record of edible mushroom cultivation.

Species	Date first cultivated (Est.)	Earliest record	Source
Agaricus bisporus	1600	1650	DeBonnefons (cited by Atkins 1979)
Auricularia auricula	600	659	So Jing (= So Gung) 659
Flammulina velutipes	800-900	Late T'ang Dynasty (618-907)	Han O (as interpreted by Zhang Shou-Cheng (1981)
Lentinus edodes	1000-1100	1313	Wang Cheng (as interpreted by Zhang Shou-Cheng 1981)
Pleurotus ostreatus	1900	1910	Falck (cited by Zadrazil 1978)
	1930's		Nie 1983
Pleurotus sajor-caju	1974	1974	Jandaik 1974
Tremella fuciformis	1800	1866	Hupei Fung-Hsien Chih (cited by Chen Sze-Yue 1983)
Volvariella volvacea	1700	1822	Yuen Yuen 1822

Source: Chang & Miles, 1987

Quimio, *et al.*, 1990), it can also be a highly industrialized agricultural enterprise with a considerable capital outlay as in the cultivation of the *Agaricus* mushroom (Chang and Hayes, 1978; Flegg *et al.*, 1985). Today, the aim of mushroom growers and researchers is to increase the yield from a given surface area, to shorten the cropping period, and to achieve a high number of "flushes" (rhythmic cycles of fruit body production) each with a high yield. To accomplish maximum output requires an understanding of substrate materials and their preparation, selection of suitable media for spawn making, the preservation of cultures, the breeding of high yield and improved quality strains, as well as improvements in the management of mushroom beds, including mushroom pest and disease control (Chang, 1991).

Mushroom science is the study of the principles and practices of mushroom cultivation. Like any branch of science, it needs systematic investigation to establish facts and principles for future development. Moreover, constant production of successful crops requires both practical experience and scientific knowledge. Practical experience can be obtained through a period of personal participation that includes training in and observation of the practices of mushroom cultivation. The scientific knowledge basically comes from mushroom biology and the related fields of science (Chang & Miles, 1989).

6. CONCLUSION

In spite of the many conceptual and technical problems still existing in the cultivation of many mushroom species, a more important role for mushrooms as a source of food protein to enrich human diet in those regions of the world where the shortage of protein is most marked is foreseen. The introduction of new technology for 1) the collection and preservation of mushroom germplasm, 2) for the breeding of strains of high yield and good quality, and 3) the maximization of mushroom production per unit area at minimum cost, will provide an even cheaper source of food protein from agricultural and industrial organic wastes. This kind of bioconversion technology is increasingly attractive and is also a continuing challenge. Many of the multiple beneficial roles of mushrooms have been shown by scientific studies. Mushroom cultivation provides mushroom protein for humans as well as spent compost for animal feed. This spent compost may also be used as a fertilizer or soil conditioner in agriculture. Mushroom cultivation counteracts deterioration of the environment by the breakdown of lignocellulosic wastes which commonly act as pollutants, thus facilitating waste management.

In addition, mushrooms have been traditionally used in China and Japan for their medicinal and tonic properties. Several pharmaceuticals have been developed from mushrooms in Japan and their active components identified (Pai *et al.*, 1990). Cosmetic products and some healthful beverages have also been produced in China from mushrooms of *Ganoderma*.

The term mushroom biology refers to the discipline that is concerned with the scientific study of mushrooms (macrofungi with distinctive fruiting bodies). It includes all of the above activities as well as the various biological subdisciplines, such as genetics, taxonomy, physiology, etc. It is believed that the term mushroom biology will bring together many diverse studies thus facilitating the dissemination of knowledge about mushrooms and greater recognition of this field of science which is increasingly affecting peoples' lives.

REFERENCES

ATKINS, F.C. (1979). Research and the mushroom grower. *Mushroom Science* **10**, 7-13.

ATKINSON, G.F. (1961). *Mushrooms (Edible and Otherwise)*. New York: Hafner Publishing Company.

CHANG, S.T. (1980). Cultivation of *Volvariella* mushrooms in Southeast Asia. *Mushroom Newsletter for the Tropics* **1**, 5-10.

CHANG, S.T. (1990). Mushroom as food. *Food Laboratory News* **21**, 7-8.

CHANG, S.T. (1991). Cultivated mushrooms. In *Handbook of Applied Mycology: Foods and Feeds*, pp. 221-240. Edited by D.K. Arora, K.G. Mukerji & E.H. Marth. New York: Marcel Dekker, Inc.

CHANG, S.T. & HAYES, W.A. (1978). *The Biology and Cultivation of Edible Mushrooms*. New York: Academic Press.

CHANG, S.T. & MILES, P.G. (1982). Introduction to mushroom science. In *Tropical Mushrooms - Biological Nature and Cultivation Methods*, pp.3-10. Edited by S.T. Chang & T.H. Quimio. Hong Kong: The Chinese University Press.

CHANG, S.T. & MILES, P.G. (1987). Historical record of the early cultivation of *Lentinus* in China. *Mushroom Journal of the Tropics* **7**, 31-37.

CHANG, S.T. & MILES, P.G. (1989). *Edible Mushrooms and Their Cultivation*. Florida: CRC Press.

CHANG, S.T. & MILES, P.G. (1991). Recent trends in world production of cultivated mushrooms. *The Mushroom Journal*. **503**, 15-18.

CHANG, S.T. & MILES, P.G. (in press). Mushroom biology - a new discipline. *The Mycologist*. (U.K.).

CHEN, SZE-YUE. (1983). *General Account of Edible Mushrooms*. Nantong, Jangsu, China: Jangsu Science and Technology Press.

DEAK, T. (1991). Culture collection: safe guards against extinction. *Nature and Resources* **27**, 30-36.

FLEGG, P.B., SPENCER, D.M. & WOOD, D.A. (1985). *The Biology and Technology of the Cultivated Mushroom*. New York: John Wiley & Sons.

GRAY, P. (1967). *The Dictionary of the Biology Sciences*. New York: Reinhold Book Corporation.

HAWKSWORTH, D.L. (1991). The fungal dimension of biodiversity: magnitude, significance, and conservation. *Mycological Research* **95**, 641-655.

JANDAIK, C.L. (1974). Artificial cultivation of *Pleurotus sajor-caju* (Fr.) Singer. *Mushroom Journal* **22**, 405.

KENDRICK, B. (1985). *The Fifth Kingdom*. Waterloo: Mycologue Publication.

MILES, P.G. & CHANG, S.T. (1986). The collection and conservation of genes of *Lentinus*. In *Cultivating Edible Fungi*, pp. 227-233. Edited by P.J. Wuest, D.J. Royse & R.B. Beelman. Amsterdam: Elsevier.

MISH, F.C. (1988). *Webster's Ninth New Collegiate Dictionary*. Beijing: World Publishing Corporation.

NASS. (1991). National Agricultural Statistics Service. *Mushrooms*. Washington, D.C.: United States Department of Agriculture.

NIE, SAN. (1983). Oyster mushroom. *Edible Fungi* **11**, 45-46.

PAI, S.H., JONG, S.C. & LO, D.W. (1990). Usages of mushroom. *Bioindustry* **1**, 126-131.

PEGLER, D.N. (1983). *The Mitchell Beazley Pocket Guide to Mushrooms and Toadstools*. London: Mitchell Beazley Publishers Ltd.

QUIMIO, T.H., CHANG, S.T. & ROYSE, D.J. (1990). *Technical Guidelines for Mushroom Growing in the Tropics*. Rome, FAO.

SNELL, W.H. & DICK, E.A. (1971). *A Glossary of Mycology*. New York: Harvard University Press.

SO, JING (or SO GUNG). (659). *T'ang Pen Tsao.*

STORK, N.E. (1988). Insect diversity: facts, fiction and speculation. *Biological Journal of the Linnean Society* **35**, 321-337.

THOMAS, C.D. (1990). Fewer species. *Nature, London* **347**, 237.

VEDDER, P.J.C. (1978). *Modern Mushroom Growing.* Netherlands: Educaboek-Culemborg.

YUEN, YUEN. (1822). *Kwangtung Tung Chin.* (In Chinese).

ZADRAZIL, F. (1978). Cultivation of *Pleurotus*. In *The Biology and Cultivation of Edible Mushrooms*, pp. 521-557. Edited by S.T. Chang & W.A. Hayes. New York: Academic Press.

ZHANG, SHOU-CHENG. (1981). *Book of Agriculture* by Wang Cheng and (consideration of) the location of the mushroom village. *Edible Fungi* **2**, 46-47 and 38.

CHAPTER 2

CULTURE COLLECTIONS

David Smith

International Mycological Institute,
Kew, Surrey TW9 3AF, United Kingdom.

1. INTRODUCTION

The storage and maintenance of organisms in a pure, viable and stable condition is essential for their use as reference strains, both in research and industrial processes. It is necessary to keep standard inocula in order that the quality and quantity of the product is maintained. In many cases, producers will keep their own starter cultures being familiar with the methods of storage and conditions required. They establish small private collections to provide working cultures for their own use, selecting suitable methods and procedures, the majority of which are common to all collections. The aims and objectives of all culture collections are similar, the differences are usually limited to the organisms kept and the size of the operation. The methods and facilities will depend upon the resources available. There are many collections established all over the world keeping a wide range of organisms, some keeping a few special strains, others keeping several thousands and making them available without restriction. The latter are sometimes termed service culture collections and may offer many other related services. It is essential that organisms supplied are correct and retain their characteristics. This requires sound quality control measures. To help culture collections set and maintain high standards, the World Federation for Culture Collections (WFCC) have produced guidelines which outline the necessary requirements (Hawksworth, 1990). The first service culture collection was that of Frantisek Kral in the German Technical University in Prague established in 1890 (Sly & Kirsop, 1990). Since that time over 350 collections have registered with the World Data Center (WDC; Takishima *et al*, 1990). These are supplemented by many private and commercial collections who do not make a commitment to supply their strains. However, this already well established genetic resource falls short of representing the vast number of organisms in nature.

Not all these collections supply fungi and even fewer supply cultures of edible fungi. Hawksworth (1985) lists 73 culture collections that include fungi; 37 hold Basidiomycetes although of these, 18 hold 100 strains or less. There are other sources of information on collections which hold strains of edible fungi, for example, the World Data Centre, Life Science Research Information Section, RIKEN, 2-1 Hirosawa, Wako, Saitama 351-01, Japan and the Information Centre for European Culture Collections (ICECC), Mascheroder Weg 1b, D-3300 Braunschweig, Germany. Table 1 also lists some of the collections or suppliers that can provide cultures of edible fungi.

In this chapter, it is the intention to discuss the basic principles involved in culture collections and to draw attention to further sources of information. The recipient of fungus cultures must be sure that the organism is correct, pure and viable and therefore must be assured that the supplier operates to appropriate high standards. The observance of good laboratory practice, much of which is outlined in the WFCC guidelines, is necessary to ensure a good quality service. Registration of collections with the WFCC ensures the practice of some of these guidelines. The stability of the host organization and its funding must be established. Important production and patent strains should remain available once deposited in a service collection and therefore it should not depend upon short term funding. Collections should have clear objectives balancing their size and the organisms held with the resources, facilities and experience available. The acquisition of too many strains would lead to storage capacity and resources being over-stretched. Collections should not aim to duplicate what is already available. Although a degree of duplication is useful, particularly for important strains, collections should aim to acquire a unique set of organisms. The correct level of staffing is also critical to a well-run collection, not only to deal with the routine acquisitions, preservation and maintenance but also to provide the appropriate level of supply and other services offered. The selection of the preservation techniques

Table 1. Some culture collections and suppliers of edible fungi.

Country	Supplier
Austria	Raiffeisen-Bioforschung GmbH Reitherstr. 21-23 A-3430 Tulln
Germany	FAL-Bundesforschungsanstalt für Landwirtschaft Institute für Bodenbiologie Bundesallee 50 W-3300 Braunschweig
	Friedrich-Schiller-Universität Jena Sektion Biologie Pilzkulturensammlung Freiherr-vom-Stein-Allee 2 0-5300 Weimar
	Versuchsanstalt für Pilzanbau der Landwirtschaftskammer Rheinland Hüttenallee 235 W-4150 Krefeld-Großhüttenhof
India	Maharashtra Association for the Cultivation of Science Department of Mycology and Plant Pathology, MACS Research Institute, Low College Road, Pune 411004
Japan	Forestry and Forest Products Research Institute 1 Matsumosato Kukizaki-machi Inashiki-gun, Ibaragi-ken 305
	The Mushroom Research Institute of Japan 8-1 Hirai-machi Kiryu-shi 376
	Tottori Mycological Institute 211 Kokoge Tottori-shi 689-11

Table 1 continued.

Country	Supplier
Netherlands	Centraalbureau voor Schimmelcultures Oosterstraat 1, PO Box 273, 3740 AG Baarn
Phillippines	Mycological Herbarium, University of the Phillippines at Los Banos, College, Laguna
UK	Horticultural Research Institute Worthing Road, Littlehampton, West Sussex BN17 6LP
USA	American Type Culture Collection 12301 Park Lawn Drive, Rockville, Maryland 20852
	Fungi Perfecti, PO Box 7634, Olympia WA 98507

for the organisms not only depends upon the success of the method but also upon the use of the organism, time, facilities and resources available. Despite the choice, at least two methods should be selected in order to guard against the loss of a strain during storage. Long-term stability should be considered together with the required availability of the culture. If strains have to be available without delay, then a collection may select a continuous growth method but should always back this up with one that reduces the possibility of change during storage. Growth techniques allow strain drift; the synthetic medium and conditions provided places selective pressure on the organism, allowing those variants growing best to dominate. Desiccation, for example freeze-drying, and freezing offer methods that can prevent such changes. However, in general it is only the fungus spore, or other structures that are produced by the organism to withstand desiccation, that survive freeze-drying. Freezing or cryopreservation is more widely applicable and methods can be developed for optimum survival of individual strains. Where cryopreservation is the only technique that allows the organism to survive, this should be backed up by alternative methods. Storage in or above liquid nitrogen can be backed up by storage in freezers that can maintain temperatures of -140°C or below. The risk of the loss of a collection is further reduced if a duplicate collection is stored in a separate building perhaps on the same site, at a distance, or in another part of the company. If this is not possible, then the deposit of important strains in a service culture collection is a useful alternative.

Culture collections must be able to ensure that an organism is correctly named. The supply of a wrongly named organism can lead to the invalidation of research project results, be time wasting and expensive. Culture identification should be validated by a specialist and if one is not available within the collection itself outside help should be sought. It is also important that names are correct so that relevant regulations can be applied. The distribution of certain human, animal or plant pathogens is restricted and it is important that culture collection staff are aware of current legislation in these matters. Quarantine and Health and Safety regulations may require containment of strains within the laboratory and, just as importantly, during transit by freight or by mail. It is necessary that relevant import permits are obtained so that unnecessary delay in transit, which may result in the loss of the organisms, can be avoided.

It is also essential that information concerning strains are adequately and correctly recorded. A unique collection number should be given to a new strain and never re-used if that particular strain is lost. Details recorded should include, the source of the isolate, substrate, host, geographical location, who identified it, special uses of the strain, conditions of growth and how it is preserved. Other information such as the organism's hazard status, the need for permits and distribution restrictions are also necessary. The name of the organism implies many properties and any deviation from the Type specimen or expected properties should also be recorded. Computerization of this data is recommended to facilitate access. The vast amount of commercially available hardware and software can cause confusion when selecting a system. There are already several national, regional or international databases in existence and the expertise in developing them should not be forgotten. An attempt has been made to develop a common format for databases for storage of culture collection data. A project sponsored by the Commission of the European Community (CEC), Biotechnological Action Programme (BAP), the Microbial Information Network Europe (MINE), have published the format used in the production of their European wide strain database for fungi (Gams *et al.* 1988) and bacteria (Stalpers *et al.* 1990). The Microbial Strain Data Network (MSDN), and WDC are further examples. Information on these, the existing culture collections and the strains held, or additional information on the above points can be obtained from the service culture collections themselves or organizations such as ICECC or WFCC.

2. PRESERVATION OF EDIBLE FUNGI

The long-term availability and stability of fungus strains depends upon the selection and application of appropriate techniques. Strains can be grown under a variety of conditions that can reduce the need for transfer. For example, limited nutrients, preventing desiccation and lowering temperature can all assist in this aim. Drying or cryopreservation can avoid the need to continually grow the organism and therefore significantly reduce the costs and time needed for maintenance. However, there are many organisms that are sensitive to such preservation techniques. Growth under special conditions can induce mutation or select variant strains and this can be exacerbated by contamination during transfer or storage or by selection from an atypical area of the colony by an inexperienced worker. The added problem that may be encountered when growing cultures on agar is the infestation by mites or insects. They cause problems by eating the fungus and bringing contaminants into culture vessels. Good laboratory hygiene and the use of physical barriers can reduce this problem significantly. Although Petri dishes can be difficult to protect, cigarette seals placed on tube or universal bottle cultures can prevent infestation. The seals are prepared by cutting cigarette papers in half, sterilizing them for 2h at +180°C and attaching them to the neck of the bottles or tubes with copper sulphate gelatine glue (20g gelatine, 2g copper sulphate in 100 ml distilled water). Excess paper is burnt away allowing caps to be replaced. The seal allows air to pass but prevents mites from entering. Acaricides, such as actelic (West Care Group, Aldershot, UK), can be used to clean laboratories and some storage conditions such as desiccation, storage under oil, freezing to -20°C or below or drying techniques can be used to reduce or prevent infestation. Wherever it is possible, collected samples and specimens should be kept away from stored or growing cultures. A separate room should be available to keep potentially contaminated material away

Table 2. Preservation of Edible and Related Fungi at IMI

Name	N° of strains in collection	Preservation Technique and number of strains preserved					
		N° of strains	Oil storage Survival* (yrs)	N° of strains	Liquid nitrogen Survival* (yrs)	N° of strains	Freeze-dried Survival* (yrs)
Armillaria mellea	48	48	22-36	13	15-20	2	2 & 7
Flammulina velutipes	1	1	18	1	18	1	18
Ganoderma applanatum	2	2	20 & 39	1	20	-	-
Ganoderma lucidum	4	4	20-39	2	20	-	-
Ganoderma miniatocinctum	4	4	1	1	1	-	-
Ganoderma philippii	2	2	5 & 27	-	-	-	-
Ganoderma tornaturm	1	1	1	1	1	-	-
Ganoderma sp.	3	3	1	-	-	-	-
Lentinus degeneri	1	1	27	1	14	-	-
Marasmius kroumironsis	1	1	11	1	11	-	-
Marasmius palmivorus	3	3	24-27	3	14	-	-
Pleurotus cystidiosus	1	1	30	1	14	-	-
Pleurotus ostreatus	3	3	1-22	1	19	-	-
Pleurotus sojor-caju	1	1	15	1	15	-	-
Volvariella esculenta	2	2	8 & 13	-	-	-	-

* Still viable in storage

from the culture collection.

It is apparent that continuous growth of fungi is not satisfactory and only of limited use for edible fungi in culture collections. Some techniques for extending the period between subcultures have been used extensively for example, storage of cultures under a layer of mineral oil or storage in sterile water. Drying techniques are useful for harvested spores and would include storage in desiccators, on silica gel, in soil, L-drying and freeze-drying. Freezing of cultures has been the most successful of all methods although viability, longevity and stability depend upon the correct preparation of the fungus, the cryoprotectant used, cooling rate and storage temperature. Recovery also depends upon the rate of thawing. Several reviews of techniques have been carried out (Jong, 1978; Smith, 1988, 1991).

At IMI the most used technique for the storage of the edible and related fungi has been under a layer of mineral oil. The length of successful storage can be as long as 39 years (Table 2). However, many strains have deteriorated or changed. By far the most successful method has been liquid nitrogen storage where 13 strains of *Armillaria mellea* have been stored for between 15 and 20 years and remain alive. The only edible fungus to fail to survive in liquid nitrogen at IMI so far has been a strain of *Volvariella volvacea.* Conversely many strains have failed to survive freeze-drying. The majority of these fungi at IMI are only mycelial and therefore are not expected to survive this technique. However, *Armillaria mellea* (2 strains) and *Flammulina velutipes* (1 strain) have survived a centrifugal freeze-drying technique (Table 2).

2.1. Mineral Oil Storage

This is a very simple method of storage that will retain viability of fungi for many years but places strains under selective pressure because of the special conditions of storage. Cultures are usually grown on short slopes of agar (30° to the horizontal) and are covered to a maximum depth of 10mm with sterile medicinal quality liquid paraffin (specific gravity 0.83 to 0.89). The paraffin is sterilized by autoclaving twice at 121°C for 15 min leaving the oil to cool and settle for 24h between each procedure. Storage in the laboratory at temperatures a little below those for optimum growth is normally selected although cultures are often placed in the refrigerator. Storage at low temperature is not always successful and before this is done the lower temperature limit for growth should be determined for the organisms to be stored. Several fungi have remained viable without subculture for over 40 years. However, several may lose vital characteristics such as the ability to develop the sexual state, various structures and certain biochemical properties.

Kobayashi (1984) recommends subculture periods from 8 to 10 years whereas Li & Chen (1981) recommended transferring the organisms following periods of 6 to 8 years growth. The organisms maintained were strains of *Agaricus campestris, Armillaria mellea, Flammulina velutipes, Lentinus edodes, Pleurotus ostreatus* and *Volvariella volvacea.* However, *L. edodes* was unable to produce fruiting bodies following 7 years of storage.

2.2. Water Storage

This technique may allow growth depending on the method adopted. One procedure is to cut agar plugs from the edges of actively growing cultures and placing them in sterile distilled water in screw cap bottles. The nutrients available in the agar will allow growth until oxygen is depleted in the storage container. Significantly less growth, if indeed any, will occur if the sterile water is inoculated with harvested spores or mycelium without growth medium. Many fungi have survived this method for 5-7 years, although loss of viability and properties can be encountered (Smith, 1991).

2.3. Soil Storage

Inoculation of spores or mycelium suspended in 1 ml of sterile distilled water into sterile soil of approximately 20% moisture content by dry weight can provide a method of storage that retains viability for 10 to 20 years. The initial growth period may allow variation and change resulting in the loss of properties. Cultures in soil stored at room temperature or in the refrigerator (4 to 7°C) can provide a good source of readily available inocula. All that is required for recovery are normal growth conditions and media. Although mycelium will survive in soil this method is not ideal for those fungi belonging to the Basidiomycota.

2.4. Silica Gel Storage

Drying in silica gel is only suitable for the fungus spore which can remain viable for periods up to and over 20 years. A suspension of the fungus is prepared in cold 5% (w/v) skimmed milk; 1 ml is added to 10g of sterile non-indicator silica gel crystals in glass universal bottles whilst in an ice bath to the depth of the silica gel crystals at approximately -20°C. The bottles are removed from the ice bath after inoculation and the crystals agitated to coat them with the suspension. The bottles are then incubated for 7-14 days to dry, the caps are then firmly screwed down and stored in an air tight container at +4°C. The cultures are recovered by sprinkling a few crystals onto a suitable agar growth medium and incubated for the required growth period. This method is recommended as a suitable alternative to freeze-drying for many fungi (Smith, 1989).

At least two bottles of each strain should be preserved when using the methods of oil, water, soil or silica gel storage. There is a possibility of contamination when recovering strains from the storage containers. Therefore, a bottle should be used to subculture from whilst another is kept solely to use when representation becomes necessary. This may be required when the length of storage becomes longer than that recommended, or cultures become contaminated, or due to the deterioration or death of the first culture.

2.5. Freeze-drying

Freeze-drying entails the freezing of the organism and its desiccation by the sublimation of ice under reduced pressure. The actual process is generally dictated by the equipment used; the rate of cooling and of drying and residual moisture content are all influenced by the machinery. The survival of fungi following freeze-drying depends upon the preparation of the inoculum, the cryoprotectant, cooling rate, temperature of drying and its rate, the residual water content, storage conditions and finally the rehydration procedure. The method provides an ideal means of keeping the organism stable for many years. Although samples deteriorate chemically in the long-term because of the presence of minute amounts of water and oxygen, heat sealed glass ampoules prevent contamination and mite infestation. The sealed ampoule or vial can be despatched immediately on request and preserved organisms will survive delays that may be encountered in transit.

2.5.1. Sample preparation. Fungus suspensions are usually prepared in a suspending medium that will offer some protection during freezing and drying. Many chemicals or mixtures of chemicals have been tried. The most common one used for fungi is skimmed milk; at IMI, a mixture of 10% (w/v) skimmed milk and 5% (w/v) inositol has been successful. Spores or other desiccation resistant structures are brought up into suspension by gentle agitation or scraping of the surface of a colony. Aliquots of 0.2 to 0.5 ml are added to ampoules or vials, these are covered to prevent contamination and either loaded directly into the freeze-drier or precooled prior to loading.

2.5.2. Freezing. The cooling of the fungi is quite critical. Centrifugal or spin freeze-driers cool by evaporation. The ampoules are placed in the centrifuge rack and spun while the chamber is evacuated. Water vapour evaporating from the large surface area of the suspension 'wedge' diffuses to the desiccant trap or condenser. This evaporation cools the suspension which eventually freezes. The diffusion of water vapour continues as the water is absorbed by the desiccant or frozen on to the condenser. The water vapour pressure in the condenser must remain below that at the surface of the suspension in order that freeze-drying continues. Ensuring that the temperature of the condenser is at least 20°C below the sample will do this. The small amount of liquid and the rapid rate of evaporation ensures a fairly fast cooling rate in the centrifugal freeze-drier. (> -10°C/min). A second type of freeze-drier relies upon freezing before evacuation either having an integral cooling method or relying upon freezing in a separate cooler. In this case cooling rates can be changed to allow the optimum rate for survival of the organism.

2.5.3. Drying. It is essential that the organism and suspension remain frozen during drying or the benefits of the method are lost. When the ice evaporates it does so leaving pores in the dried material where the ice crystals were. This allows further evaporation to take place and the ice interface to recede through the suspension. If the suspension is allowed to melt the system collapses and the structure is lost, possibly impeding further drying. The temperature of the suspension should ideally be below -15°C and it should not be allowed to rise above this until the residual moisture content is 5% by dry weight or less. The final moisture content should not fall below 1% and be ca 2% (Smith, 1986). If the suspension of the organism is not dried sufficiently, viability will be high initially but the suspension will deteriorate rapidly. *Armillaria mellea* survives freeze-drying but if the residual moisture content is 10% or above by dry-weight it fails to recover after 1 year storage whereas viable cells can be recovered after several years when the water content is between 1 and 2%.

Over drying is thought to cause mutations and in the fungi the lower limit is thought to be 1% by dry weight. The addition of a suspending medium, particularly a sugar, can act as a water buffer and prevent over drying. Ampoules or vials should be sealed to prevent leakage of air or moisture into the dry suspension. Back filling the ampoules with a dry inert gas such as argon or nitrogen or sealing them under vacuum will reduce potential deterioration. The rate of deterioration can be reduced by storing the ampoules at low temperature.

2.5.4. Recovery. Rehydration of the fungi should be carried out slowly giving time for absorption of moisture before culturing on a suitable medium. If the dried fungus is resuspended in water and it is immediately plated onto agar the water may be preferentially absorbed by the agar and the organisms may not be rehydrated. The following procedure is recommended for samples prepared in single neutral glass ampoules.

(i) Open ampoules by scoring the tube midway down the length of the cotton wool plug with a diamond scribing point or ampoule file and cracking the ampoule by touching it on the score with the tip of a hot glass rod.

(ii) Reconstitute the dried suspension by adding three or four drops of sterile distilled water from a Pasteur pipette and allowing 30 min for absorption of the moisture. Take care not to create aerosols during the rehydration process.

(iii) Streak the contents of the ampoule onto a suitable agar medium with an inoculating loop, and incubate at an appropriate growth temperature.

Suppliers of freeze-dried cultures will provide instructions on how to rehydrate samples. These may differ from collection to collection but must be followed to get optimum results. Vials with rubber bungs are much easier to open by removing the metal seal and the rubber bung before adding the required amount of liquid and replacing the bungs for rehydration. It is important to check the viability of stored ampoules frequently (every 1-2 years), at least until it is confirmed that the number of viable cells remain high.

Although many of the edible fungi tend to produce only mycelium in culture, the spores of strains can be harvested from mature sporophores in the production line. These harvested spores survive the freeze-drying process extremely well. Where stroma or other mycelial structures are formed these may survive the process, for example with the honey fungus, *Armillaria mellea*. Rhizomorph initials survive a shelf freeze-drying method at IMI (Smith, 1986) where the cooling rate was -1°C/min and they were dried to between 1 and 2% residual water content by dry weight. Although the freeze-drying of hyphae is not very successful, research is continuing at IMI and the Centraalbureau voor Schimmelcultures (CBS), The Netherlands, in particular. The optimization of cooling and drying protocol and the selection of suitable protective media seem to be the key to the improvement of this technique.

2.6. Cryopreservation

The storage at ultra low temperatures has proved to be the most successful method for the retention of both the viability and characteristics of fungi. The method has been used for many years. Hwang (1960) adapted a technique used by Polge *et al.* (1949) for the preservation of avian spermatozoa and many culture collections still use a similar method today. The early work was extremely successful; therefore, very few studies were carried out to determine the effect of freezing and thawing on filamentous fungi. Mazur (1968) extended our knowledge but few examples of fungi were examined. The vast number of species that are recorded (>69 000) and the even greater number there are estimated to be (1.5 million species) in the yet little explored ecological niches of the world (Hawksworth, 1991) requires extensive study to determine optimum storage conditions. More recent work by Coulson *et al.* (1986), Smith *et al.* (1986) and Morris *et al.* (1988) has shown that the initial method of preservation was not optimum and theories on the response of fungus cells to freezing and thawing not correct for all species. This coupled with the improvement to the practical aspects of cryopreservation (Challen & Elliott, 1986; Stalpers *et al.* 1987) and the availability of improved equipment have meant that vast improvements have been made to the cryopreservation technique for fungi. No one procedure will be optimum for all strains although the aspects affecting survival are generally common. Sample preparation, cryoprotectant (suspending medium), cooling rate, storage temperature and thawing rate will all affect the viability and, in many cases, the stability of the recovered strain.

2.6.1. Sample preparation. The fungus is grown on a suitable growth medium and usually suspensions of spores, fungal hyphae or plugs from colonies are taken to provide the inoculum for freezing. However, several cultures can be damaged and predisposed to further injury during the freezing procedures. For most mycelial organisms, cutting agar plugs from growing cultures is the preferred method, transferring the plugs to a suitable amount of cryoprotectant in ampoules, cryotubes or straws. To reduce mechanical damage to those strains that are more susceptible to injury, cultures can be grown on small amounts of agar in cryotubes and when mature the cryoprotectant is added to the tube prior to freezing. An alternative method is to grow the organism in liquid culture and either to mix this with equal quantities of double strength cryoprotectant or filter out the fungus and place it in cryoprotectant. If none of these are suitable, the prepared cut plugs can be allowed to repair by continuing incubation under normal growth conditions overnight before freezing.

2.6.2. Cryoprotectant. The organism must be allowed to acclimatize in the cryoprotectant chosen, usually for 1-2h. This will allow the low molecular weight chemical to penetrate the cell and replace a proportion of the cytoplasmic water. Many chemicals or chemical mixtures have been used to prevent injury during freezing and thawing. The two most commonly used for fungi are 10% (v/v) glycerol or 10% (v/v) dimethyl sulphoxide (DMSO) both of which have proven beneficial to many organisms (Smith, 1983; Challen & Elliott, 1986). Other cryoprotectants that have been tried at IMI are trehalose, proline, polyvinyl pyrollidone (PVP) and mixtures of glucose and DMSO (Smith, 1983). Although one cryoprotectant may be more effective for some organisms than other chemicals it is a combination of cyroprotectant and cooling rate that most effects the recovery of strains. DMSO is more effective at fast rates of cooling than glycerol. However, at slower rates similar high viabilities may be found in glycerol. Glycerol has, however, failed to protect *Volvariella volvacea* at IMI (Morris *et al.* 1988) whilst other workers have found that DMSO is more effective for this species (Challen & Elliott, 1986). This has also been found to be true for *Agaricus xanthadermus* and *Lepista nudum* (Challen & Elliott, 1986). The mechanism of protection is not fully understood but it is considered, in simple terms, that the lower molecular weight compounds penetrate the cells and replace water which is lost during exosmosis when external ice is formed. The presence of the cryoprotectant within the cell when internal freezing occurs reduces the ice crystal size and thus reduces mechanical damage. The larger molecular weight compounds, for example PVP, do not penetrate the cell and protect by reducing extracellular ice contact with the cell. Further details on the effect of cryoprotectants is given elsewhere (Grout & Morris, 1988).

Several fungi have been found to recover from freezing without a cryoprotectant. The zygomycetes, *Mortierella elongata* and *Mucor racemosus*, the ascomycete, *Sordaria fimicola* and the basidiomycetes, *Lentinus edodes* and *Schizophyllum commune* all recover when frozen in dilute growth medium. However, it is the hyphomycetes that show the greatest resistance to freezing without a cryoprotectant. The addition of 10% (v/v) glycerol generally improves viability dramatically. In the case of *L. edodes* it allows a further 70% of propagules to survive (Morris *et al.* 1988). In some cases, the addition of cryoprotectant chemicals to suspending media can prove deterimental. The recovery and infectivity of *Puccinia abrupta* var. *partheniicola* urediniospores is adversely affected by the addition of 10% solutions of glycerol, DMSO, trehalose or PVP in water (Holden & Smith, 1992). Careful consideration should be given to the selection of the suspending medium for cryopreservation. For best results the response of each individual strain of the organism should be studied and protocols developed for each. However, this is generally impractical and routinely either 10% (v/v) glycerol or 10% (v/v) DMSO should be selected for edible fungi.

2.6.3. Cooling rate. Generally a cooling rate of -1°C/min is employed for fungi prior to their storage at low temperature (Smith & Onions, 1983). However, more recent studies have revealed that cooling rates giving optimum recovery differ from fungus to fungus (Smith *et al.* 1986; Morris *et al.* 1988). The hyphae of *Serpula lacrimans* give highest recovery after cooling at -0.5°C/min in 10% v/v glycerol. Other fungi require much faster rates to give greatest survival, e.g. *Wallemia sebi* -77°C/min and a species of *Aureobasidium*, from rocks in Antarctica, -200°C/min, again in the cryoprotectant glycerol. However, many edible fungi survive well when cooled at slow rates of cooling. Many workers suspend cultures in the neck of nitrogen refrigerators where the evolving cold nitrogen gas cools the ampoules, straws or vials and their contents. Inevitably the cooling achieved is not linear. Cooling begins quickly slowing down as the fungus suspension temperature gets closer to that of the nitrogen gas. The cooling curve is further disrupted when latent heat is released during the freezing of the water and again when the eutectic point, the freezing point of the remaining concentrated solution is reached. Although such a cooling procedure can be successful and can be reproduced from batch to batch, it is more desirable to use a programmable freezer which will enable linear cooling at different rates depending on the range of fungi to be frozen. Alternatively, ampoules can be immersed in cold alcohol baths (Morris & Farrant, 1972). *Lentinus edodes*, -1 to -3.5°C/min, and *Volvariella volvacea*, -1°C/min, are examples of the edible fungi that survive optimally at slow rates of cooling.

2.6.4. Storage temperature. Once frozen the fungi must be stored either in a freezer or in a liquid nitrogen refrigerator. There are a wide range of storage temperatures used from -20°C to -196°C. However, at storage temperatures of -70°C and above, chemical and biochemical reactions can occur albeit slow at the lower temperature. At temperatures below this down to -140°C, physical changes in the structure of water can occur which results in the increase of ice crystal size. These changes may cause damage to the stored organism and such injury may result in the death, deterioration or a change in the organism's characteristics. A study sponsored by the CEC under BAP involving 6 European laboratories concerned the improvement of preservation techniques for biotechnological important strains. The storage temperatures of -20, -40, -80, -135 and -196°C were compared. Not only did some strains fail to recover from -20 or -40°C after only 1 or 2 years storage but those that survived showed morphological and physiological changes (Smith *et al.* 1990a, 1990b). To avoid these changes, storage temperatures must be held below -140°C at all times. There are freezers that can operate at -150°C and storage in liquid nitrogen at -196°C can ensure this. However, storage in the vapour phase above liquid nitrogen can not guarantee constant temperatures. It is only when storage is in intimate contact with the liquid itself that low temperature is retained. When the ampoules, tubes or straws are placed in the metal drawers of a metal inventory control system and the drawer rack partially immersed in liquid nitrogen temperatures are maintained below -180°C. The BAP project showed that there was little difference in recovery and stability of the strains stored at -80, -135 and in or above liquid nitrogen in the short term (2-3 years). However, it is envisaged that in the long term changes may occur in strains stored at temperatures above -140°C.

2.6.5. Thawing rate. The rate of warming from the frozen state affects the viability of fungi. *Penicillium expansum* gave highest viability following rapid warming after fast cooling to -196°C (Smith *et al.* 1986). Rates of ca +200°C/min are achieved when the ampoules are immersed in a water bath at +35°C or placed in the chamber of a programmable cooler on a thawing programme. This fast rate of warming is usually required for optimum recovery of most fungi. Slow rates of warming can allow the growth of ice crystals which may cause structural damage before the organism is fully thawed. Following thawing, the organism must be placed on an appropriate recovery medium and incubated under suitable conditions. Placing an organism onto a limiting medium immediately after thawing may not allow it to grow. After thawing there is quite often a delay in recovery when the organism is undergoing repair of damage incurred during freezing and thawing. The delay in recovery reflects the extent of the damage.

2.6.6. Development of improved cryopreservation techniques. The use of cryogenic light microscopy allows the observation of fungi during freezing and thawing. A conduction stage allows the accurate control of temperature via a computer controlled stage heater whilst the stage is cooled by the passage of cold nitrogen gas through it. This equipment allows the response of cells to freezing and thawing to be observed and images can be recorded on video tape. Fungi have been seen to shrink at slow rates of cooling and in some cases intracellular ice has been observed at faster rates (Morris *et al.* 1988). The IMI facility is being used to look at a wide range of fungi from many different environments. Fungi respond differently at different cooling rates; of the first 20 strains examined, 14 different optimum rates were recorded (Morris *et al.* 1988). Observations made can demonstrate the optimum protocol in a very short time where viability studies can take many weeks. This is especially important when patent deposits are made requiring safe preservation for the long-term. If the organism is new and its response to freezing is unknown, this could take some time. The use of the cryogenic microscope can provide the necessary information and within 24h it is possible, in most cases, to have the organism preserved. The equipment also provides information on the response of the organisms, as well as being a practical aid. The response of *Lentinus edodes* has been studied in detail. Although no intracellular ice has been observed at cooling rates up to -100°C/min, extensive shrinkage can be observed at all rates of cooling. Glycerol reduces this effect at slow rates of cooling but at fast rates shrinkage occurs even in the presence of glycerol. Roquebert (1992) has studied the same strain of *L. edodes* using transmission electron microscopy and has found that membrane material and cell cytoplasm are lost during shrinkage which is not re-absorbed during thawing. The cell does not reach its original volume on thawing and membrane material and cytoplasm is found between the cell protoplast and the cell wall. Similar occurrences have been observed in the nuclear membrane with the possibility of more serious consequences. Following membrane deletion the cell does not attain its original volume but usually recovers and grows normally in culture.

Such work is essential to help develop good preservation protocols as cell viability following recovery does not always give a true reflection of a successful method. If the resulting organism is abnormal as a result then this is not satisfactory. If the organism fails to produce the characteristics required of it, this has obvious consequences. However, if the changes are not initially apparent the problem may not manifest itself until it is too late.

3. SUMMARY

Culture collections provide backup to the every day needs of industry, research and teaching establishments. Not only do they supply strains when required but can also provide many other services. They must be able to identify the strains they hold so many can offer identification services. They offer training in culture collection techniques and management, provide a bank or reserve collection for safe deposit and duplicate research or process strains. Many operate as International Depository Authorities (IDA's) and accept patent strains under the Budapest Treaty. They have expertise in the preservation of organisms and this should be utilized.

It has been the intention here to introduce the activities of culture collections, give information on how to find those that supply edible fungi and methods that can be used to preserve these organisms. The WFCC have produced a series of publications on culture collections, one in particular gives background information on fungal collections (Hawksworth & Kirsop, 1988). The WDC and ICECC can offer information on culture collections worldwide; contact them or the nearest service culture collection for further information.

REFERENCES

CHALLEN, M.P. & ELLIOTT, T.J. (1986). Polypropylene straw ampoules for the storage of microorganisms in liquid nitrogen. *Journal of Microbiological Methods* **5**, 11-23.

COULSON, G.E., MORRIS, G.J. & SMITH, D. (1986). A cryomicroscopic study of *Penicillium expansum* hyphae during freezing and thawing. *Journal of General Microbiology* **132**, 183-190.

GAMS, W., HENNEBERT, G.L., STALPERS, J.A., JANSENS, D., SCHIPPER, M.A.A., SMITH, J., YARROW, D. & HAWKSWORTH, D.L. (1988). Structuring strain data for the storage and retrieval of information on fungi and yeasts in MINE, Microbial Information Network Europe. *Journal of General Microbiology* **134**, 1667-1689.

GROUT, B.W.W. & MORRIS, G.J. (1988). *The effects of low temperature in biological systems*. London: Edward Arnold.

HOLDEN, A. & SMITH, D. (1992). Effect of cryopreservation methods in liquid nitrogen on viability of *Puccinia abrupta* Diet & Holw var. *partheniicola* urediniospores. *Mycological Research* **96**, 91-163.

HAWKSWORTH, D.L. (1985). Fungus culture collections as a biotechnological resource. *Biotechnological and Genetic Engineering Reviews* **3**, 417-453.

HAWKSWORTH, D.L. & KIRSOP, B.E. (1988). *Living Resources for Biotechnology: Filamentous Fungi*. Cambridge: Cambridge University Press.

HAWKSWORTH, D.L. (1990). *Guidelines for the Establishment and Operation of Collections of Cultures of Microorganisms.* WFCC Secretariat, Brazil: WFCC.

HAWKSWORTH, D.L. (1991). The fungal dimension of biodiversity; magnitude, significance and conservation. *Mycological Research* **95**, 641-655.

HWANG, S. -W. (1960). Effects of ultralow temperature on the viability of selected fungus strains. *Mycologia* **52**, 527-529.

JONG, S.C. (1978). Conservation of cultures. In *The Biology and Cultivation of Edible Mushrooms*, Edited by S.T. Chang & W.A. Hayes. London & New York: Academic Press.

KOBAYASHI, T. (1984). Maintaining cultures of Basidiomycetes by mineral oil method I. *Bulletin of Forestry and Forestry Products Research Institute* **325**, 141-147.

LI, Z.Q. & CHEN, Y.Y. (1981). An evaluation of mineral oil seal preservation of basidiomycetes cultures. *Acta Microbiogica Sinica* **21**, 45-52.

MAZUR, P. (1968). Survival of fungi after freezing and desiccation. In *The Fungi: III. The Fungal Population*, pp. 325-394. Edited by G.C. Ainsworth & A.S. Sussman. New York and London: Academic Press.

MORRIS, G.J. & FARRANT, J. (1972). Interactions of cooling rate and protective additive on the survival of washed human erythrocytes. *Cryobiology* **9**, 173-181.

MORRIS, G.J., SMITH, D. & COULSON, G.E. (1988). A comparative study of the morphology of hyphae during freezing with the viability upon thawing of 20 species of fungi. *Journal of General Microbiology* **134**, 2897-2906.

POLGE, C., SMITH, A.U. & PARKES, S. (1949). Revival of Spermatozoa after dehydration at low temperatures. *Nature, London* **164**, 666.

ROQUEBERT, M.F. (1992). Freezing of *Lentinus edodes*. *Mycological Research*, in the press.

SLY, L.I. & KIRSOP, B. (1990). 100 years of culture collections. *Proceedings of the Kral Symposium to Celebrate the Centenary of the First Recorded Service Culture Collection*. Osaka, Japan: Institute of Fermentation.

SMITH, D. (1983). Cryoprotectants and the cryopreservation of fungi. *Transactions of the British Mycological Society* **80**, 360-363.

SMITH, D. (1986). *The evaluation and development of techniques for the preservation of living fungi.* Ph.D. thesis, University of London.

SMITH, D. (1988). Culture and Maintenance. In *Living Resources for Biotechnology: Filamentous fungi*, pp. 75-199. Edited by D.L. Hawksworth & B.E. Kirsop. Cambridge: Cambridge University Press.

SMITH, D. (1989). Techniques used for the preservation of viability and stability of fungi. *Review of Tropical Plant Pathology* **6**, 1-26.

SMITH, D. (1991). Maintenance of filamentous fungi. In *Maintenance of Microorganisms and Cultured Cells*, pp133-159. Edited by B.E. Kirsop

& A. Doyle. London: Academic Press.

SMITH, D., COULSON, G.E. & MORRIS, G.J. (1986). A comparative study of the morphology and viability of hyphae of *Penicillium expansum* and *Phytophthora nicotianae* during freezing and thawing. *Journal of General Microbiology* **132**, 2014-2021.

SMITH, D. & ONIONS A.H.S. (1983). *The Preservation and Maintenance of Living Fungi.* Kew: CAB Mycological Institute.

SMITH, D., TINIGNER, N., HENNEBERT, G.L., de BIEVRE, C., ROQUEBERT, M.F. & STALPERS, J.A. (1990). Improvement of preservation techniques for fungi of biotechnological importance. In *Biotechnology R & D in the EC I, Catalogue of Biotechnological Action Programme BAP Achievements*, pp. 115-117. Edited by A. Vassariott & E. Mangien. Paris: Elsevier.

SMITH, D., TINIGNER, N., HENNEBERT, G.L., de BIEVRE, C., ROQUEBERT, M.F. & STALPERS, J.A. (1990b). Improvement of preservation techniques for fungi of biotechnological importance. In *Biotechnology in the EC II. Detailed Final Report of Biotechnological Action Programme BAP Contractors*, pp. 89-94. Edited by A. Vassarotti & E. Magnien. Paris: Elsevier.

STALPERS, J.A., DEHOOG, A. & VLUG, I.J. (1987). Improvement of the straw technique for the preservation of fungi in liquid nitrogen. *Mycologia* **79**, 82-89.

STALPERS, J.A., KRACHT, M., JANSENS, D., DE LEY, J., VAN Der TOORN, J., SMITH, J., CLAUS, D. & HIPPE, H. (1990). Structuring strain data for storage and retrieval of information on bacteria in MINE, Microbial Information Network Europe. *Systematic and Applied Microbiology* **13**, 92-103.

TAKISHIMA, Y., SHIMUIRA, T., UDAGAWA, Y. & SUGAWARA, H. (1990). *Guide to World Data Center of Microorganisms with a List of Culture Collections in the World*, 249pp. Samitama: World Data Center of Microorganisms.

CHAPTER 3

BIOLOGICAL BACKGROUND FOR MUSHROOM BREEDING

Philip G. Miles

Department of Biological Sciences, State University of New York at Buffalo, Buffalo, New York 14260, U.S.A.

1. INTRODUCTION

The essential feature of any breeding programme is the bringing together of desired traits possessed by two different individuals. Breeding programmes will also include the creation and selection of desired traits, but it is the assembling in one individual stock of the best combination of genetic material for the production of mushrooms of high quality and yield that is the goal of the mushroom breeder. To accomplish this the breeder must have a thorough knowledge of the basic biology and breeding system of the mushroom species which he is trying to improve. In this chapter the basic biology of fungi will be examined with an emphasis upon cultivated edible species.

2. GROWING FUNGI IN CULTURE

2.1. Obtaining Cultures

In order to do mushroom breeding the breeder must be able to grow the fungus in culture - a fungus that he or someone else originally obtained from nature. This culture may be obtained **from tissue** of a mushroom collected in the field. Tissue taken from the upper portion of the stipe, or from the pileus, or even from immature gills, commonly serves as the starting material for cultures. The surface area of the mushroom may be wiped clean with an antiseptic solution and then the outermost layer of tissue removed by cutting it away with a sterile scalpel, or the inner tissue may be exposed by simply breaking the cap open. Finally, a piece of the exposed inner tissue is transferred aseptically to an agar medium that supports the growth of fungi. Even when the operation has been imperfectly performed and some bacterial growth occurs, the fungal mycelium may grow through the agar and leave the bacteria behind. Thus, it is not difficult to obtain pure cultures from the tissue of mushrooms. Another way to obtain cultures is **from spores**. Spores which have been discharged over a sterile agar medium, or spores which have been discharged over a sterile plate, such as the inside surface of the lid of a petri dish, can be picked up aseptically in sterile distilled water, diluted, and eventually plated onto an agar surface for germination. The spores, or germlings arising from the spores, may be isolated singly or in mass to establish cultures.

In general, it is preferred to use tissue cultures to establish the working material for a breeding program, for one can have confidence that this material from the mushroom had the genetic capability of forming a mushroom, whereas the genetic competence of single spore isolates, or multispore cultures, is unknown. This is a generalization to which there are some exceptions. For subsequent breeding studies, however, the establishment of single spore isolates is essential.

3. SPORE GERMINATION

Spore germination is sometimes easily accomplished, and sometimes only with difficulty. For genetic studies it is important that germination of spores should occur in high percentage so that there will not be interference with genetic ratios or the elimination of a class of isolates being studied due to such things as linkage to lethal factors.

The conversion of a spore from an inactive state to an actively growing condition, eventually leading to the formation of hyphae, is what we mean by spore germination. The general requirements for germination of viable spores include: adequate availability of moisture, an adequate supply of oxygen, a suitable temperature, and a suitable pH. For a particular species it must be determined what is "adequate" or "suitable". If these requirements are proper for a species, the spores of some species will germinate immediately upon being released from the parent structure.

3.1. Dormancy

There are, more commonly, species whose spores remain dormant for a period of time. This dormancy is of two types - endogenous (also called constitutive or constitutional) and exogenous. In **endogenous dormancy** the spores do not germinate even under environmental conditions which are favourable for subsequent growth. In **exogenous dormancy** environmental factors delay germination.

3.2. Breaking Dormancy

Removal from this dormant stage to an active stage may be accomplished in various ways for different species. The spore walls may be relatively impermeable to water and to gases, and this may keep the spore in the dormant stage unless the outer spore wall is removed or rendered permeable. Activation may be brought about by temperature shock as has been well documented for *Neurospora tetrasperma*. There are also self-inhibitors of spore germination which have been studied in detail in specific species. These self-inhibitors of germination may be volatile or non-volatile substances, the removal of which must be accomplished if germination is to take place. Self-inhibitors have been found to operate in different ways (e.g., inhibition of RNA translation and of protein activation, and enzyme inactivation). In some cases, a specific compound has been identified as the inhibitor (e.g., methyl-cis-3,4 dimethoxycinnamate in rust uredospores).

Chemical compounds which activate spores (stimulate germination) are also well known. In the presence of furfural there is increased respiratory activity (O_2 consumption and CO_2 evolution) promoting spore germination. Furfural is effective for *Neurospora* and *Coprinus radiatus*. Also, a number of compounds which disrupt the lipoprotein membranes and thus increase permeability have been demonstrated to be effective in stimulating germination.

There are fungal species whose spores require certain nutrients for germination. Such **exogenously dormant species** may show increased germination in the presence of specific nutrients which include such things as glucose and minerals (including trace elements). Members of the genus *Agaricus* were considered for a long time to have spores that were difficult to germinate - a factor that retarded breeding for strain improvement. Observations that isolated spores seldom germinated but that much better germination occurred when spores were close together suggested that gaseous substances stimulated germination, and it was subsequently discovered that spore germination increased when the spores were placed in the same gaseous environment of mycelia of *Agaricus* or other fungi.

Found to be operative here was isovaleric acid, a volatile substance produced by the metabolizing mycelium. Isovalerate removes the carbon dioxide which is a self-inhibitor of germination. This self-inhibition of germination by carbon dioxide occurs by virtue of the fact that CO_2 is normally fixed to form oxaloacetate. The production of oxaloacetate in the spores suppresses the activity of the enzyme succinic dehydrogenase, thus slowing down the respiratory activities of the TCA cycle in the spores with the consequence that the spores remain dormant. Isovalerate, as indicated, removes the carbon dioxide, because it is a precursor of a CO_2 - acceptor, β-methylcrotonyl coenzyme A. This alternative route for CO_2 utilization prevents formation of more oxaloacetate so that there is sufficient respiratory activity of the TCA cycle to permit germination. Spores of *Volvariella bombycina* may have a similar mode of germination (Chiu & Chang, 1987).

3.3. Composition of Medium

From the foregoing, the composition of the medium upon which the spores are sown for germination prior to isolation of germlings is a consideration. The meiospores of most species possess endogenous nutrients which are sufficient for the transition to the germling stage when environmental conditions are suitable. Such spores will germinate on water agar which may be advantageous to the investigator, for in such a non-nutrient medium the growth of germlings is less than that on a nutrient medium and there is less opportunity for the mycelia of neighbouring germlings to mix before single spore isolation is attempted. There are cases in which some supplementation may enhance germination and some cases in which a medium component, such as the ammonium ion, may suppress the growth of isolated germlings

(Shaw & Miles, 1970; Fries, 1966). The optimal conditions for spore germination and subsequent growth of germlings must be determined for the species being investigated and even for stocks within the species.

3.4. Morphology of the Germinating Spore

Morphologically, the first sign of spore germination is a uniform swelling of the spore. Some spores have a pore through which the germ tube emerges. In any event there is a polar growth involving movement of vesicles to a location that becomes the site of formation of the germ tube. With the insertion of wall components at this site, an apex has developed that constitutes the tip of the germ tube. Upon further growth the germ tube becomes a young hypha. The hypha is a tubular structure with a cell wall consisting primarily of polysaccharides (e.g., glucans and chitin), and on the inside of the wall is the cell membrane which forms the outside limits of the cytoplasm. Various cytoplasmic inclusions are present - endoplasmic reticulum, Golgi bodies, mitochondria, vacuoles, various types of vesicles, and nuclei. Studies with the electron microscope have clearly shown that these various structures are bounded by membranes.

4. MYCELIAL GROWTH

4.1. The Hyphal Apex

The apex of the hypha contains large numbers of vesicles and has been termed the **AVC** (apical vesicular complex) by Burnett (1976), with other organelles being located farther back from the apex. An outstanding feature of hyphal growth is that it takes place at the tip (apical growth). The hyphae of the fungi that produce mushrooms are septate. At intervals, cross walls develop along the hypha, dividing the hypha into units which are loosely referred to as cells.

4.2. Nutritional Requirements

The nutritional requirements for mycelial growth are relatively simple. Since the fungi are heterotrophic organisms, they must be supplied with a source of **carbon**. While many carbon sources may be used, individual species commonly have a preference, but among the simple sugars glucose is most frequently preferred, and in amounts of approximately 2%. In addition to the simple sugars, polysaccharides may provide carbon for the fungi. Such polysaccharides are the usual source of carbon for fungi in nature, and the insoluble polysaccharides, such as cellulose, are broken down by extracellular enzymes to simpler, soluble units which are then taken into the fungal hyphae by **absorptive (osmotrophic) nutrition**. Under certain conditions a number of organic compounds (alcohols, organic acids, polycyclic compounds, and amino acids) may also provide carbon for mycelial growth. It is worth mentioning at this point that a mixture of sugars may give greater growth than simply the summation of growth to be obtained by each separately (Horr, 1936). On the other hand, another frequent observation in studies of carbon nutrition is that when a fungus is supplied with a mixture of carbon sources, it may use one preferentially over the others. The matter of concentration of carbon source is also important in determination of the effectiveness of promotion of growth. This has been shown in *Coprinus lagopus* (= *C. cinereus*) by Moore (1969) with the demonstration that growth on sucrose is negligible at low concentrations but occurs at higher sugar concentrations.

Obviously, all organic compounds (carbohydrates, amino acids, lipids, nucleic acids) require carbon in their skeletal framework but it should not be overlooked that the carbon compounds supplied to fungi also provide the energy required for the organism's metabolic activities.

Nitrogen is a required element in media used for the growth of fungi. It is essential for the synthesis of fungal proteins, purines, pyrimidines, and is also necessary for the production of chitin, a common fungal cell wall polysaccharide that is composed of units of N-acetylglucosamine. While there are a few fungi that have been reported in the past to fix atmospheric nitrogen, there is no confirmation, using modern techniques, that this is true and there certainly are no filamentous fungi that do. To date, nitrogen fixation is known to occur only in prokaryotic organisms. Thus, the common sources of nitrogen in fungal media are salts of nitrate and ammonium, and organic nitrogen compounds. A generalization can be made to the effect that the nitrogen requirements of all fungi can be met by organic nitrogen (e.g., peptone or amino acids), some may utilize the ammonium ion, and some may use nitrates. Those that utilize nitrate are also able to use the ammonium ion. In the cell the ammonium ion is combined with α-ketoglutaric acid in the presence of glutamic dehydrogenase to form glutamic acid, and other amino acids may be formed by transaminase reactions. Thus, there is a relationship between ammonia and TCA cycle intermediates which leads to the formation of amino acids.

A medium for the growth of fungi must contain **minerals**. The mineral requirements are similar to those for plants. While some fungi require a reduced form of **sulfur**, most species utilize sulfur as sulfate (e.g., magnesium sulfate) in a range of 0.0001 to 0.0006 M. The role of sulfur is for sulfur-containing amino acids (e.g., cysteine and methionine), for vitamins such as thiamine and biotin, and in some cases for products of secondary metabolism (e.g., penicillin). **Phosphorus** is present in ATP, nucleic acids, and the phospholipids of membranes. It is commonly included in growth media as potassium phosphate at a concentration of about 0.004 M. **Potassium** has the role of a cofactor in many enzyme systems and its requirement is fulfilled at a concentration of 0.001 to 0.004 M.

Many enzymes are activated by **magnesium**. essential to all fungi, and magnesium sulfate, when supplied at a concentration of 0.001M, satisfies this requirement.

Equally important mineral elements, although required in lower concentrations, are the **trace elements**: iron, zinc, manganese, copper, and molybdenum. These are constituent elements in enzymes and are not all universally required by fungi.

Vitamins are organic molecules required in small amounts and not used as a source of energy or structural material of protoplasm. The vitamin has a catalytic action and imparts specificity in its function as a coenzyme. The vitamin requirement is influenced by temperature and pH since it is concerned with enzyme activity. Most fungi are able to make their own vitamins, but sometimes in amounts too low to give optimal growth. **Thiamine** (vitamin B_1) is a natural deficiency of a number of basidiomycetes, including the wood-rotting edible mushrooms *Lentinus edodes* and *Flammulina velutipes*. Biotin (vitamin B_7 or vitamin H) is a natural deficiency for some fungi such as the ascomycetes *Neurospora* and *Sordaria*.

A chemically defined medium that supports the growth of many edible basidiomycetes is as follows:

Dextrose	20.0 g
Asparagine	2.0 g
KH_2PO_4	0.46 g

K_2HPO_4	1.0 g
$MgSO_4.7H_2O$	0.5 g
Thiamine-HCl	0.12 mg
Distilled H_20	1000 ml

4.3. Physical Factors

The physical factors for growth of mycelium may differ from those for fruiting. These physical factors act in conjunction with the nutritional requirements for growth. Such factors are commonly presented as the cardinal points - minimum, optimum, and maximum. In growth of fungi, the physical factors greatly affecting growth are temperature, light, moisture, and aeration.

While the **temperature** extremes are of great importance in determining the survival and distribution of fungal species in nature, it is the effect of temperature upon enzyme activities that is of greatest interest to the experimentalist and to the mushroom grower. In the linear phase of growth for each 10° C increase in temperature the growth rates double (i.e., the Q_{10} is 2). Obviously, this cannot go on indefinitely, for high temperatures inactivate enzymes. In some cases, it has been shown that the failure to grow at higher temperatures was the result of inability to synthesize a required vitamin, and growth of the fungus would take place at the higher temperature if that vitamin were supplied in the medium.

Strong **light** may inhibit mycelial growth or even kill the fungus, although the growth of the mycelia of most fungi is not sensitive to light. There are reports that the inhibition by strong light has been reversed by the addition of natural materials (containing vitamins) to the medium. Thus, the effect of light may have been on the light destruction of vitamins formed by the fungus. Phototropic responses of reproductive structures of fungi are well known and much studied, but of greatest interest for the general topic of this book will be the effect of light upon the development of fruiting body primordia and stages of fruiting body development. These will be considered later on in this chapter.

Most fungi require high **moisture** levels. For most Basidiomycetes, maximum growth is obtained with a relative humidity of 95 to 100%. A moisture content of about 50 to 75% was found by Flegg (1962) to be optimum for the growth of mushroom mycelium, and the maintenance of a high relative humidity of the air in the mushroom houses reduces evaporation from the substrate surface.

Oxygen and **carbon dioxide** are the components of air that are of importance to the fungi. Oxygen is important because the edible fungi carry on aerobic respiration. While concentrations of carbon dioxide greater than 0.3 to 0.5% typically result in inhibition of the formation of fruiting body primordia and to promote stipe elongation of the mushrooms that develop (Flegg *et al.*, 1985), mycelial growth is increased by concentrations of 0.1 to 0.5% CO_2 (San Antonio & Thomas , 1972).

5. SEXUALITY

So far we have dealt primarily with the mycelial or vegetative stage of the life cycle of a mushroom. This is of great importance in mushroom production, but for breeding work it is essential to understand the sexual reproductive phase. Sexuality may conveniently be thought of as consisting of three cardinal events - plasmogamy, karyogamy, and meiosis. **Plasmogamy** is the fusion of protoplasts, as a consequence of which different nuclei are brought into the same cell. **Karyogamy** is the fusion of nuclei, and **meiosis** is the reductional division of the diploid nucleus which is formed by karyogamy, resulting in the formation of four haploid nuclei.

5.1. Events Leading to Karyogamy

5.1.1. Plasmogamy. First, we shall examine the events leading to karyogamy as found in Basidiomycetes, since most of the edible fungi are in this class of fungi. In this class, plasmogamy occurs by the fusion of hyphae or sometimes by fusion of a hypha and a spore. Hyphal fusions occur in different ways. Buller (1933) described three types of **hyphal fusion**: tip-to-tip, peg-to-peg, and tip- to-peg. Ahmad and Miles (1970) could not always distinguish the peg in the tip-to-peg type of fusion, and classified these as tip-to-side. The tip-to-side was the most numerous type of hyphal fusion observed in *S. commune*. In each case, fusion occurs between actively growing segments of both hyphae; that is, fusion involves the growing hyphal apices and thus all fusions are really end-to-end fusions as Buller pointed out over fifty years ago. The pegs that grow out are essentially lateral branches, each with a growing tip.

It has been demonstrated in some species that the frequency of hyphal fusions is dependent upon mating type considerations. In *S. commune* (Ahmad & Miles, 1970) and in *Coprinus* (Smythe, 1973), higher hyphal fusion frequency was associated with confrontations between strains in which there was a common allele of one of the mating type factors. This effect of higher fusion frequency was shown to occur when the confronted strains were separated by a permeable membrane, indicating the activity of a diffusible substance. Voorhees and Peterson (1986) have shown a chemotropic **attraction of hyphae to basidiospores** of *S. commune* and that the attractant is produced by spores only in the presence of a mycelium (although the mycelium may be separated from the spores by a permeable membrane). Only viable spores will produce the attractant which is no longer produced after fusion has occurred.

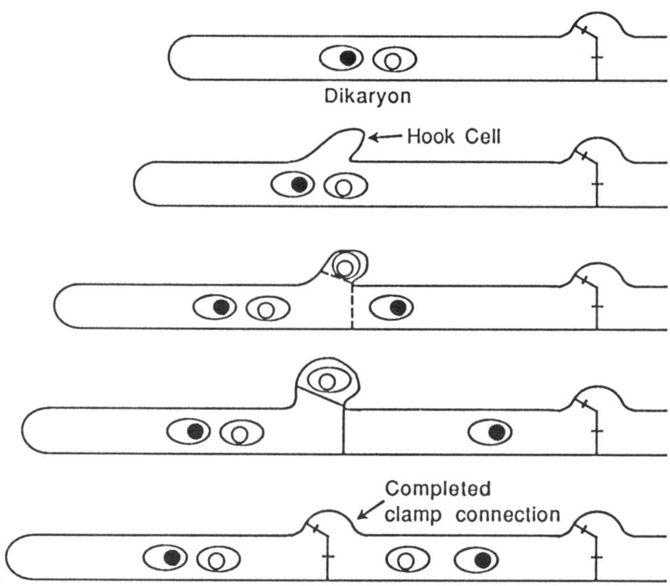

FIGURE 1. Clamp Connection Formation. Compatible nuclei undergoing simultaneous division. The hook cell provides a temporary location for one of the daughter nuclei so that the dikaryotic condition can be maintained in the apical cell.

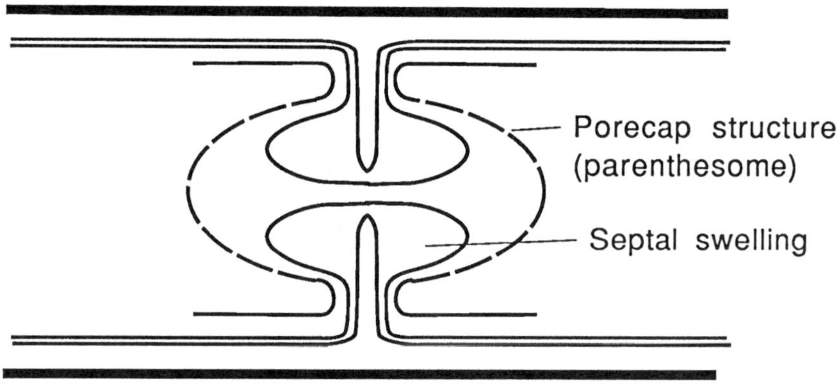

FIGURE 2. Cell-to-cell movement of organelles is inhibited by the pore cap and septal swellings.

There have been numerous cases in which hyphae have been observed to curve and grow toward one another, but the identity of the causal factor has not been made (Ahmad & Miles, 1970). Roles of chemicals in the development of fruiting bodies of the Basidiomycetes will be considered briefly later in this chapter.

5.2. Nuclear Migration

In many Basidiomycetes a phenomenon occurs following hyphal fusion which permits the establishment of a heterokaryotic condition in growing cells. This phenomenon is **nuclear migration** by which is meant the migration of nuclei through the established mycelium of a confronting strain. Nuclear migration was first described by Buller from observations of the formation of clamp connections, indicative of dikaryotic hyphae, at a distance from the site of fusion of compatible strains greater than could be explained by hyphal growth. Nuclear migration rates of different species may vary widely from 0.5mm/h in *Coprinus lagopus* (Buller, 1931) to 40mm/h in *Coprinus congregatus* (Ross, 1976). The advantage to the organism of nuclear migration is that the heterokaryotic condition is not limited to those heterokaryotic cells that have been formed by plasmogamy, but can be established rapidly in large numbers of cells and be perpetuated in the growing cells as is the case in the clamp forming dikaryotic hyphae of Basidiomycetes (Figure 1). With the discovery of the dolipore septum (Figure 2) of basidiomycetes, a difficulty in understanding nuclear migration arose. How does the relatively large nucleus pass through the small constrictions of the dolipore septum? At least one mechanism has been demonstrated to occur, and this is probably the principal one. Giesey and Day (1965) first found broken down septa of *Coprinus* in electron micrographs of hyphae through which nuclei were believed to have migrated. Mayfield (1974), working with a mutant of *S. commune* whose individual hyphae could be singled out for fixation and sectioning for electron microscope studies, showed the dissolution of the dolipore septum as nuclei advanced toward the septum during nuclear migration. The control of nuclear migration by the *B* mating type locus has been demonstrated in *S. commune* and mating type control of nuclear migration in other tetrapolar Basidiomycetes may be of common occurrence.

5.3. Heterokaryosis

In most cases in nature it is the heterokaryotic mycelium that ultimately develops into the fruiting body. The secondary mycelium of basidiomycetes which forms dikaryotic hyphae by means of clamp connection formation constitutes the best example of this. The edible mushrooms *Lentinus*, *Flammulina*, *Pholiota*, *Pleurotus*, *Tremella*, *Dictyophora*, *Auricularia*, *Hericium*, and *Coprinus* are dikaryotic heterokaryons. Here the dikaryotic condition with clamp connections is formed by the mating of compatible monokaryotic strains. Following hyphal fusion and nuclear migration, the dikaryotic mycelium is formed. Thus, a microscopic examination revealing clamp connections is evidence for the heterokaryotic condition in which there is a 1:1 ratio of the two nuclear types.

5.3.1. Recognition of heterokaryosis or heterozygosis. The recognition of heterokaryosis is not always so simple. In *Agaricus bitorquis* it has been shown by Raper (1976) that the fertile heterokaryon formed by the mating of two compatible homokaryotic strains forms no clamp connections although it is dikaryotic. **Macroscopic observation** reveals heavier growth at the line of interaction of the confrontation between compatible strains. The aerial hyphae in the line of interaction are revealed by phase contrast microscopy or by staining and microscopic examination to be binucleate (actually dikaryotic since it is fertile). Confirmation of the heterokaryotic condition of the line of interaction was also obtained by **complementation of auxotrophic strains** (Raper, 1976). With heterokaryosis only at the line of interaction, it is apparent that nuclear migration does not occur and that the dikaryons develop from hyphal fusion and the subsequent proliferation by cellular division. The entire mycelium of neither of the confronting mycelia becomes heterokaryotized.

In the formation of heterokaryons in the ascomycete *Morchella* (the edible morel), a similar line of interaction occurs. This was originally called a "barrage" (Hervey, *et al.*, 1978) as being similar to the barrage reaction in some Basidiomycetes, but in this barrage reaction there is an area of sparse heterokaryotic growth between the confronting homokaryotic mycelia. In *Morchella* there is a line of heavy growth between the confronting mycelia that I refer to as an "overlap", since its appearance is similar to that reaction in *Cyathus stercoreus* as described by Fulton (1950). In their demonstration of heterokaryosis in *Morchella*, in which they used markers of **mutants resistant to chemical inhibitors of growth** on selective media to confirm heterokaryosis, Volk and Leonard (1989) termed the area of heavy growth between the confronting mycelia as a "meld". The role of the heterokaryon in the life cycle of the morel has not yet been completely elucidated, but as in the case with most fungi, it probably plays an important role in fructification.

Armillaria mellea remained a mystery for a long time as far as its mode of sexuality is concerned. The studies of Ullrich and Anderson (1978) and others on **nuclear ploidy** have revealed the presence of diploid nuclei in the vegetative hyphae. This is a bifactorial heterothallic fungus with multiple alleles at both loci. Clamp connections are not formed but the diploid mycelium obtained from tissues of fruiting bodies, or as a result of compatible matings, display a depressed, crustose mycelium as opposed to the fluffy, aerial mycelia of monosporous isolates - permitting macroscopic identification of a compatible mating. The demonstration of prototrophy following the mating of auxotrophic strains confirmed the heterozygotic condition. The diploid nature of the nuclei of vegetative mycelia and of mycelia formed following mating of compatible homokaryotic strains, and the haploid nature of nuclei of monosporous mycelia, were determined using mithramycin, a fluorescent antibiotic that complexes with DNA. Measurements of nuclear fluorescence were then made (Franklin *et al.*, 1983).

Absolute proof of heterokaryosis, with the exception of clamp forming dikaryotic hyphae, requires genetic complementation tests with strains bearing different markers so that the heterokaryon alone grows on selective media. These may be auxotrophic markers in which only the heterokaryon grows on a minimal medium, or they may be drug resistant markers in which only the heterokaryon (composed of resistant strains) will grow on a medium containing the chemicals in question. There is a distinct advantage in the latter system that stems from the fact that it is easier to select for chemically resistant mutants than it is for auxotrophic mutants which are commonly recessive.

5.4. Development of the Fruiting Body

Information on the chemical nature of the fungal cell wall is important for 2 reasons: 1) development and morphogenetic alterations of the fruiting bodies are the consequences of changes in cell wall composition; and 2) to obtain protoplasts for breeding purposes the cell wall must be enzymatically degraded. While there is much variation in the chemical composition of the cell walls of various species of fungi, qualitatively it can be said that the cell walls of most edible mushrooms are made up principally of chitin and glucans. Thus, enzymes that will break down chitin and glucans are used to obtain protoplasts. Fortunately, for the mushroom breeder wishing to employ protoplast fusion techniques, there are a number of commercial enzyme preparations from which he can select those which work best for the species with which he is working.

Mushrooms are macrofungi with distinctive fruiting bodies. This structure is large enough to be used as a source of food, and obtaining a high yield of mushrooms of good flavour and texture is the objective of the mushroom grower. The scientist is thus concerned with the development of the mushroom and with the genetic, nutritional, physical, and chemical factors that influence its development. The term mushroom to most people brings to mind the image of the common button mushroom or champignon (*Agaricus bisporus*) with its **cap**, **gills**, and **stipe**. There are, however, other types of fruiting bodies that are commercially cultivated.

5.4.1. Stages of development. With many distinctive types of mushrooms, it is difficult to generalize on stages of development. In mushroom development the grower thinks in terms of **vegetative mycelial growth** (spawn running or mycelial running), **pinning** (production of mushroom primordia), **button stage** (membrane closed and stipe still short), and **caps** with well developed membranes and slightly longer stipes. The researcher studying development uses more precise definitions, for other researchers must be able to apply the same criteria to their studies. It is almost necessary to establish distinct criteria for different genera.

One of the organisms in which biochemical studies of morphogenesis of fruiting have been made is the wood-rotting basidiomycete *S. commune*, a popular experimental fungus. The early studies of Wessels (1965) raised many questions whose solutions called for synchronous development of fruiting bodies since assays required more material than was available in one small fruiting body. Thus, the developmental stages described by Leonard and Dick (1968) came to be used by other workers (e.g., Schwalb, 1971). There are five stages following the vegetative mycelial stage in this system. Stage I appears as a knot of interwined hyphae, the primordium. In Stage II there is a stalk that forms, and in Stage III an apical pit, within which can be seen microscopically the hymenium and gills. Stages IV and V involve growth of the differentiated fruiting body, with continued development of the basidia and spore formation following the nuclear events of karyogamy and meiosis.

A second basidiomycete which has been much investigated as to development of the fruiting body is *C. cinereus*. From the days of Brefeld (1877), and accepted for decades, was the concept that the multicellular fruiting body developed from a single cell. Experimental work has shown that in *Coprinus* this is not the case but that the primordium comes from hyphal aggregates (Matthews & Niederpruem, 1972).

Much of the recent work on morphogenesis of fruiting has dealt with the biochemical events accompanying fruiting body formation. The details of these studies are beyond the scope of this article but there are excellent reviews that may be consulted (Gooday, 1982; Moore, 1988; Schwalb, 1978).

6. CONTROL OF FRUITING

6.1. Genetic (Other Than Mating Type)

Control of fruiting may have a genetic basis in addition to the steps in differentiation regulated by mating type gene activity that have been the subject of morphogenetic investigations. In *S. commune*, it was shown by Raper and Krongelb (1958) that dikaryons derived from an extensive sample of homokaryons, worldwide in origin, varied widely in their ability to form fruiting bodies under standard conditions of nutrition and environment. The results obtained from studying the ability to fruit of 3100 dikaryons indicated that fruiting competence was genetically inherited and that component strains within dikaryons that displayed "good fruiting" were dominant to strains that displayed "poor fruiting". That is, any mating involving a "good fruiting" strain would fruit early and well, but a dikaryon derived from two "poor fruiting" strains would fruit late and slightly or not at all. The genetic control of fruiting is polygenic.

This control of fruiting competence by genes other than the mating type genes may be of general occurrence in Basidiomycetes, but this has not been investigated. There is some evidence in *L. edodes* that this is the case based upon differences in primordia formation and fruiting body development in liquid culture by ten different dikaryotic stocks (Miles & Chang, 1987). The genetic control of fruiting body morphology by a number of mutants of different species is well documented.

6.2. Monokaryotic

Fruiting by monokaryotic strains has been studied in *Polyporus ciliatus* by Esser and Stahl (1975). They have indicated that **monokaryotic fruiting** results from the presence of specific alleles at two loci. The allele fi+ initiates monokaryotic fruiting and, in the presence of fi+, fb+ leads to the production of small, fertile fruiting bodies. Interestingly, 60% of the monosporous isolates form monokaryotic fruiting bodies. A similar situation in regard to the number of monokaryotic fruiters exists in *Pholiota nameko*. Arita (1979) reported cases in which 70% of the monosporous isolates of this unifactorial (bipolar) heterothallic fungus produce fruiting bodies. The phenomenon of monokaryotic fruiting has also been reported in *Coprinus*, *Flammulina*, *Pleurotus*, and *Schizophyllum*, and future studies may find a role for this information in mushroom breeding.

6.3. Nutritional Requirements

Both nutritional and environmental factors may differ widely for mycelial growth and for fruiting. As a generalization, conditions that favour vegetative growth are not satisfactory for fruiting. It is frequently observed that a high concentration of nutrients encourages vegetative growth and that fruiting may be stimulated by mechanisms that cause the cessation of vegetative growth. The nature of the **carbon** source also influences fruiting with certain sources producing abnormal fruiting bodies. The most commonly used carbon source in laboratory studies is glucose, and there are contradicting reports as to whether fruiting is inhibited by continuous supplies of glucose.

In **nitrogen** matters the situation in regard to fruiting is similar to that for carbon in that species differ in their ability to utilize nitrogen sources. The concentration of nitrogen that is best for fruiting is slightly higher than that for mycelial growth. This is ambiguous, however, for high concentrations of nitrogen encourage mycelial growth and thus decrease fruiting body formation. In fruiting, the C:N ratio is very important.

In **mineral nutrition** only calcium has been reported to give qualitative differences between vegetative growth and fruiting. In the gasteromycete *C. stercoreus*, Lu (1973) demonstrated the necessity of calcium in the medium for fruiting. It has been suggested, however, that in the matter of concentration of minerals, a concentration sufficient for vegetative growth may not be adequate for sporulation.

A situation similar to that for minerals occurs with **vitamins**. That is, the requirements for vegetative growth and fruiting body formation may be similar, but with a higher concentration required for fruiting than for vegetative growth. Thiamine (vitamin B_1) functions as cocarboxylase, the coenzyme of carboxylase whose activity is in the conversion of pyruvic acid to acetaldehyde and carbon dioxide. This regulation of carbohydrate metabolism is important in the development of fruiting bodies. The cultivated mushrooms *Auricularia*, *Coprinus*, *Flammulina*, *Lentinus*, *Pholiota*, *Pleurotus*, and *Volvariella* all have a thiamine requirement for mycelial growth and the formation of fruiting body primordia and/or fruiting bodies.

6.4. Physical Factors

In consideration of mycelial growth, it was indicated that the physical factors for growth of mycelium may differ from those for fruiting. Klebs (1900) generalized on this subject in the following way: a) external conditions determine whether growth or reproduction takes place; b) conditions which are favourable for reproduction are always less favourable for growth; c) growth may take place under a wider range of environmental conditions than reproduction, with growth taking place under conditions that inhibit reproduction; d) vegetative growth is a preliminary step that creates a suitable internal environment for reproduction. Specifically, concerning the fungi, we shall see what generalizations can be made regarding the physical factors of the environment and fruiting.

The **pH** requirements for growth and fruiting differ as to their optimal values but not necessarily in the same direction from one species to another. The pH change that occurs during growth (e.g., by the production of organic acids) may trigger the response from vegetative growth to fruiting.

Both the **temperature** range and the range of reported optimum temperatures for fruiting are narrower than they are for vegetative growth.

There are numerous fungi whose reproduction is apparently uninfluenced by **light** in the visible range. On the other hand, there are some fungi with particular light requirements for fruiting. The effects of light upon fungal development have been reviewed by Tan (1978) and only a few examples will be given here. It has been shown in *S. commune* (Bromberg & Schwalb, 1976) that light is required for formation of primordia (Stage I) and for the early stages of fruiting body development in which short cylindrical stipes with terminal apical pits are formed (Stages II and III), but it is not required for the appearance of gills, subsequent growth, and expansion to the mature fruiting body. However, the functioning of the fruiting body in production of spores ceases in the dark, with a light period of 5 to 6 hours required for recovery from this dark period of inhibition of sporulation. In a number of mushrooms (e.g., *F. velutipes* and *Polyporus brumalis*), light is required for the normal expansion of the pileus. In *Favolus arcularius*, *Panus fragilis*, and *Boletus rubinellus*, pileus formation is induced by light, but, when transferred to darkness, stipes form but no pilei (Kitamoto *et al.*, 1968, 1972, 1974a,b).

A. bisporus presents a contrasting system, for in this fungus light is actually inhibitory to the development of primordia, as well as having an inhibitory effect upon the developmental stages of stipe elongation and pileus expansion.

Phototropic responses in the stipes of many mushrooms are responsible for the alignment of gills or tubes bearing the hymenial layer so that basidiospore discharge will send the spores into positions where they will then fall free of the hymenial surface.

Mushroom growers are well aware of a need for proper **aeration** in the mushroom house. Adequate oxygen is essential for mycelial running and a slightly raised level of carbon dioxide may increase vegetative growth, but levels of 0.3 to 0.5% have an effect upon fruiting body development as well as upon primordia formation. The effect is an elongation of the stipe and at even higher concentrations the development of the pileus is affected. Experimental studies have shown that 5% CO_2 completely prevented primordium development in *S. commune* (Niederpruem, 1963).

6.5. Chemicals

Chemical substances have been reported to induce fruiting in Basidiomycetes. Such a compound is referred to as a fruiting-inducing substance (FIS). Studies by Leonard and Dick (1968) indicated that fruiting bodies of *S. commune* could be induced by cell-free extracts from *Hormodendrum*, and *A. bisporus*, as well as from *S. commune*. The studies of Uno and Ishikawa (1973) demonstrated the existence of a fruiting-inducing substance (FIS) in *C. cinereus*, and, furthermore, that the FIS is cyclic adenosine monophosphate (cyclic AMP or cAMP). The FIS of *PYS. commune* is not cyclic AMP, although cAMP does have an effect upon the development of fruiting bodies (Schwalb, 1974).

7. LIFE CYCLES

Experimental manipulation of a species for breeding purposes requires knowledge of its life cycle. There are numerous variations in life cycles and frequently some details for a particular species are uncertain. Nevertheless, it is possible to represent some general patterns, one of which may fit in major respects the species in which the breeder is interested.

7.1. Heterothallism

Sexuality in the Basidiomycetes was first described by Kniep (1920) in *S. commune* and Bensaude (1918) in *Coprinus fimetarius* based upon matings of monosporous mycelia, with the formation of clamp connections indicating compatible reactions in these heterothallic species. Following the work of Kniep and Bensaude, it was discovered that heterothallism took two major forms, bifactorial and unifactorial.

7.1.1. ***Bifactorial.*** In the form first encountered by Kniep and Bensaude, two unlinked mating type factors were found to be operative, with a heteroallelic condition at both loci required for compatibility and the formation of the dikaryotic condition. This was called tetrapolar incompatibility (tetrapolarity) or may be referred to as heterothallism with bifactorial control.

7.1.2. ***Unifactorial.*** In the other type of heterothallism there is a single mating type factor with a heteroallelic condition at that locus bringing about compatibility and the formation of the dikaryotic condition. This has been called bipolar incompatibility (bipolarity) or may be referred to as heterothallism with unifactorial control. In both systems multiple mating type alleles exist.

Heterothallism in ascomycetes such as *Saccharomyces*, *Neurospora*, and *Ascobolus* is generally found to be unifactorial with alternate mating type alleles, but there are variations that occur.

Previously we examined sexuality in the Basidiomycetes from plasmogamy through karyogamy and up to meiosis in the fruiting body. We should now look at meiosis in reference to mating type control. In the basidium, karyogamy brings together the two compatible nuclei of the dikaryon to form a diploid nucleus. It is this diploid nucleus which undergoes meiosis resulting in the formation of four haploid nuclei. Each of these nuclei moves through a short stalk (sterigma) on the basidium into the developing basidiospores. Thus, if we represent the mating type loci as *A* and *B*, the dikaryon can be symbolized as *A*x*B*x + *A*y*B*y, the diploid nucleus as *AxAy BxBy*, and the meiospores as *AxBx*, *AxBy*, *AyBx*, and *AyBy*. Since these occur in equal frequency, that is , in a ratio of 1:1:1:1, it is evident that the *A* and *B* loci are unlinked. These spores will germinate and form homokaryotic mycelia. When mycelia from spores are confronted in all possible combinations, the following results occur:

	AxBx	*AxBy*	*AyBx*	*AyBy*
AxBx	-	-	-	+
AxBy	-	-	+	-
AyBx	-	+	-	-
AyBy	+	-	-	-

+ represents the presence of clamp connections indicating a dikaryon (= two compatible nuclei in the same cell) and - represents the absence of clamp connections.

Thus, it can be seen that compatibility occurs only in those combinations that are heteroallelic for both mating type factors. The compatible reaction is the one that commonly leads through growth of dikaryotic mycelium to the development of the fruiting body. There may be a large number of alleles at both mating type loci.

In bipolar heterothallism there is only one mating type locus, commonly designated the *A* locus, and compatibility occurs when mycelia bearing different mating type alleles are brought together (e.g., *Ax* x *Ay*) producing dikaryotic hyphae (*Ax* + *Ay*) which leads to the formation of the fruiting body. Unifactorial heterothallism is present in *P. nameko*, and *A. bitorquis*.

It should be pointed out that in some species (e.g., *S. commune*, *F. velutipes*, and *C. cinereus*), it has been demonstrated that the mating type factors consist of 2 subunits located in short chromosomal segments and that crossing over can take place between the subunits. Such crossover mating types are then compatible with both parental mating types. A heteroallelic condition at one of the subunits (alpha or beta) when strains are confronted is sufficient to establish functionality of the factor. The possible occurrence of mating type subunits should always be considered in any species in which breeding is to be attempted.

7.2. Homothallism

In some species germination of a single basidiospore will lead to a mycelium that will form fruiting bodies and the completion of the life cycle. This is called homothallism and we now recognize two types of homothallism, primary and secondary.

7.2.1. Primary. Primary homothallism occurs when a homokaryotic mycelium arises from a single spore which contained a single postmeiotic nucleus. The only cultivated mushroom that is thought by some to display primary homothallism is *V. volvacea*, but uncertainty as to how variation arises among single spore isolates in successive generations indicates that details of the life cycle are unknown. Li (1991) has demonstrated by the use of strains bearing genetic markers that heterokaryons can be formed when monosporous mycelia are confronted. These heterokaryons, however, did not fruit, and thus the genetic behaviour of the markers could not be followed through the next generation.

7.2.2. Secondary. The cultivated mushroom that is produced in greatest amounts and for which there is the most advanced technology is *A. bisporus*. This is also the mushroom whose life cycle remained an enigma for a very long time. It is now generally agreed that *A. bisporus* is secondarily homothallic. A mating type system is operative that is really that of a unifactorial heterothallic species, but the 2-spored nature of the basidia conceals the heterothallism since following meiosis two nuclei enter each spore. Such spores may contain two compatible nuclei, and consequently a mycelium that is self-fertile. In secondary homothallism of *A. bisporus*, the homokaryotic phase found in the primary homothallic species is lacking, except in the uninucleate spores of the 3 and 4-spored basidia that occur only in approximately 5% of the basidia. It is known that not all of the spores produced on 2-spored basidia give rise to self-fertile mycelia. This has been explained in the following manner. If the 4 post-meiotic nuclei involved in the unifactorial mating type system are labelled *Ax1*, *Ax2*, *Ay1*, and *Ay2* and random movement of nuclei to give binucleate spores occurs, then spores of the following nuclear constitution are possible: *Ax1Ax2*, *Ax1Ay1*, *Ax1Ay2*, *Ax2Ay1*, *Ax2Ay2*, and *Ay1Ay2*. Since only those spores bearing both *Ax* and *Ay* alleles will give rise to fertile spores, it means that two- thirds of spores from 2-spored basidia would be self-fertile, and one-third would be self-sterile as suggested by Langton and Elliott (1980).

However, there is evidence that the movement of postmeiotic nuclei into basidiospores is nonrandom. By analysis of several isozyme loci, it has been possible to identify genotypic classes of *A. bisporus* (=*A. brunnescens*), and Royse and May (1982) showed that almost 90% of single-spore isolates were heteroallelic at these isozyme loci, arguing against the concept of random distribution of postmeiotic nuclei in this species. Such random distribution would expect about two-thirds of the spores to be heteroallelic.

A different experimental approach has also led to the conclusion that non-sister nuclei are preferentially incorporated into basidiospores following the second division of meiosis, i.e., that the incorporation of 2 post meiotic nuclei into a basidiospore is not a matter of random probability. The experimental procedure in these studies made use of the determination of the phenotypes of restriction fragment length polymorphisms of parental homokaryons, parental heterokaryons, and the single spore f1 progeny from fruiting bodies resulting from the mating of the parental homokaryons. For mycelium obtained from each single spore culture, DNA was extracted and subjected to restriction enzymes, with the subsequent assay for restriction fragments by Southern hybridization using plasmids carrying nuclear DNA segments from *A. bisporus* as probes.

The results (Summerbell *et al.*, 1989) were that 351 of a total of 367 (95.6%) single spore isolates were heterokaryons that had restriction fragment length polymorphisms which were identical to those of the heterokaryon formed from the homokaryotic parents. The single spores each contained 2 nuclei (a heterokaryotic condition) and those 351 single spore isolates were heteroallelic for all the restriction fragment length polymorphism loci which were heteroallelic in the parents of the heterokaryon.

8. IMPORTANCE OF EXPERIMENTAL STUDIES

The mushroom breeder needs to have up-to-date information on such things as: spore germination, the nutritional and environmental requirements for both mycelial growth and the development of the mushroom, basic and modern genetic principles and techniques, and an intimate understanding of all phases of sexuality and the life cycle of the species he wants to improve through breeding techniques. For much of this, he relies on the research of others. Some of this comes from the experience of the grower, who through observation and study, may come to an intuitive understanding of a fundamental principle about mushroom biology. The scientific basis for this understanding may come from the efforts in research of applied scientists, or it may come from basic studies of fungi, even those of different species from the one desired to be improved through breeding.

Very fruitful in the advancement of knowledge of mushroom biology has been research with the experimental organisms *Schizophyllum commune* and *Coprinus cinereus*, but edible mushrooms such as *Flammulina* (*Collybia*) *velutipes* and species of *Pleurotus* have been used in basic studies as well as research in cultivation techniques. Of course, fundamental knowledge of many aspects of fungal biology such as the effect of light upon fruiting, the role of carbon dioxide in vegetative growth and fruiting, microbial fermentation in composting, and extracellular enzymes in mushroom mycelium, to name a few, have resulted from applied studies concerned with mushroom cultivation.

The interaction of growers, applied mushroom scientists and mycologists has advanced mushroom science in the past and will stimulate further progress in mushroom breeding. It is evident that there is a tremendous opportunity to utilize knowledge gained and techniques developed in such areas as molecular genetics and protoplast fusion technology to the breeding of edible mushrooms. The biological background important to mushroom breeding is ever expanding.

REFERENCES

AHMAD, S.S. & MILES, P.G. (1970). Hyphal fusions in the wood-rotting fungus *Schizophyllum commune*. I. The effects of incompatibility factors. *Genetical Research, Cambridge* **15**, 19-28.

ARITA, I. (1979). Cytological studies on *Pholiota*. *Report of Tottori Mycological Institute* (Japan) **17**, 1-118.

BENSAUDE, M. (1918). *Recherches sur le cycle evolutif et la sexualite chez les Basidiomycetes*, Ph.D. Thesis, University of Paris, Nemours, France.

BREFELD, O. (1877). *Botanishe Untersuchungen uber Schimmelpilze*, III. Heft. *Basidiomyceten*, pp. 1- 226. Leipzig. (From Raper, 1966).

BROMBERG, S.K. & SCHWALB, M.N. (1976). Studies on basidiospore development in *Schizophyllum commune*. *Journal of General Microbiology* **96**, 409-413.

BULLER, A.H.R. (1931). *Researches on Fungi IV.* London: Longmans, Green & Company.

BULLER, A.H.R. (1933). *Researches on Fungi, V.* London: Longmans, Green & Company.

BURNETT, J.H. (1976). *Fundamentals of Mycology*, 2nd Edition. London: Edward Arnold, Ltd.

CHIU, S.W. & CHANG, S.T. (1987). The activation of spore germination in *Volvariella bombycina. Mushroom Journal of the Tropics* **7**, 61-66.

ESSER, K. & STAHL, U. (1975). A genetic correlation between dikaryotic and monokaryotic fruiting in basidiomycetes. *Proceedings of the First Intersectional Congress IAMS* **1**, 294 - 300.

FLEGG, P.B. (1962). The development of mycelial strands in relation to fruiting of the cultivated mushroom (*Agaricus bisporus*). *Mushroom Science* **5**, 300-313.

FLEGG, P.B., SPENCER, D.M. & WOOD, D.A. (1985). *The Biology and Technology of the Cultivated Mushroom.* New York: John Wiley & Sons.

FRANKLIN, A.L., FILION, W.G. & ANDERSON, J.B. (1983). Determination of nuclear DNA content in fungi using mithramycin: vegetative diploidy in *Armillaria mellea* confirmed. *Canadian Journal of Microbiology* **29**, 1179-1183.

FRIES, N. (1966). Chemical factors in the germination of spores of Basidiomycetes. In *The Fungal Spore*, pp. 189-200. Edited by M.F. Madelin. London: Butterworths.

FULTON, I.W. (1950). Unilateral nuclear migration and the interactions of haploid mycelia in the fungus *Cyathus stercoreus. Proceedings of the National Academy of Sciences of the United States of America* **36**, 306-312.

GIESEY, R.M. & DAY, P.R. (1965). The septal pores of *Coprinus lagopus* (Fr.) sensu Buller in relation to nuclear migration. *American Journal of Botany* **52**, 287-293.

GOODAY, G.W. (1982). Metabolic control of fruitbody morphogenesis in *Coprinus cinereus.* In *Basidium and Basidiocarp: Evolution, Cytology, Function, and Development*, pp. 157-173. Edited by K. Wells & E. K. Wells. New York: Springer-Verlag.

GRIFFIN, D. H. (1981). *Fungal Physiology.* New York: John Wiley & Sons.

HERVEY, A., BISTIS, G.N. & LEONG, I. (1978). Cultural studies of single ascospore isolates of *Morchella esculenta. Mycologia* **70**, 1269-1274.

HORR, W.H. (1936). Utilization of galactose by *Aspergillus niger* and *Penicillium glaucum. Plant Physiology* **11**, 81-99.

KITAMOTO, Y., TAKAHASHI, M. & KASAI, Z. (1968). Light induced formation of fruit-bodies in a basidiomycete, *Favolus arcularius* (Fr.) Ames, *Plant and Cell Physiology* **9**, 797-805.

KITAMOTO, Y., SUZUKI, A. & FURUKAWA, S. (1972). An action spectrum for light-induced primordium formation in a basidiomycete, *Favolus arcularius* (Fr.) Ames. *Plant Physiology* **49**, 338-340.

KITAMOTO, Y., HORIKOSHI, T. & KASAI, Z. (1974a). Growth of fruit-bodies in *Favolus arcularius. Botanical Magazine* (Tokyo) **87**, 41-49.

KITAMOTO, Y., HORIKOSHI, T. & SUZUKI, A. (1974b). An action spectrum for photoinduction of pileus formation in a basidiomycete, *Favolus arcularius, Planta* **119**, 81-89.

KLEBS, G. (1900). Zur Physiologic du Fortflanzung einiger Pilze. III. Allegemeine Betrachtungen. *Jahrbuch wissenschaft Botanische* **35**, 80-203. (From Lilly & Barnett, 1951, and Griffin, 1981)

KNIEP, H. (1920). Uber morphologische und physiologische Geschlechstdifferenzierung (Untersuchungen an Basidiomyzeten). *Verhandlung Physiologie Medizin Gesellschaft, Wurzburg* **46**, 1-18.

LANGTON, F.A. & ELLIOTT, T.J. (1980). Genetics of secondarily homothallic Basidiomycetes. *Heredity* **45**, 99-106.

LEONARD, T.J. & DICK, S. (1968). Chemical induction of haploid fruit bodies in *Schizophyllum commune. Proceedings of the National Academy of Sciences of the United States of America* **59**, 745-751.

LI, S. (1991). Genetical and Cytological studies on variations of *Volvariella volvacea.* Ph.D. Thesis, Hong Kong. The Chinese University of Hong Kong.

LILLY, V.G. & BARNETT, H.L. (1951). *Physiology of the Fungi.* New York: McGraw-Hill.

LU, S.H. (1973). Effect of calcium on fruiting of *Cyathus stercoreus. Mycologia* **65**, 329-334.

MATTHEWS, T. & NIEDERPRUEM, D.N. (1972). Differentiation in *Coprinus lagopus.* I. Control of fruiting and cytology of initial events. *Archives of Microbiology* **87**, 257-268.

MAYFIELD, J.E. (1974). Septal involvement in nuclear migration in *Schizophyllum commune. Archives of Microbiology* **95**, 115 - 124.

MILES, P.G. & CHANG, S.T. (1987). Fruiting of *Lentinus edodes* (Shiitake) in liquid media. *Mircen Journal of Applied Microbiology and Biotechnology* **3**, 103-112.

MOORE, D. (1969). Sources of carbon and energy used by *Coprinus lagopus* sensu Buller. *Journal of General Microbiology* **58**, 49-56.

MOORE, D. (1988). Recent developments in morphogenetic studies of higher fungi. *Mushroom Journal for the Tropics* **8**, 109-128.

NIEDERPRUEM, D.J. (1963). Role of carbon dioxide in the control of fruiting of *Schizophyllum commune*. *Journal of Bacteriology* **85**, 1300-1308.

RAPER, C.A. (1976). Sexuality and life cycle of the edible, wild *Agaricus bitorquis*. *Journal of General Microbiology* **95**, 54-66.

RAPER, J.R. (1966). *Genetics of Sexuality in Higher Fungi*. New York: Ronald Press.

RAPER, J.R. & KRONGELB, G.S. (1958). Genetic and environmental aspects of fruiting in *Schizophyllum commune* Fr. *Mycologia* **50**, 707-740.

ROSS, I.K. (1976). Nuclear migration rates in *Coprinus congregatus*; a new record? *Mycologia* **68**, 418-422.

ROYSE, D.J. & MAY, B. (1982). The use of isozyme variation to identify genotypic classes of *Agaricus brunnescens*. *Mycologia* **74**, 93-102.

SAN ANTONIO, J.P. & THOMAS, R.L. (1972). Carbon dioxide stimulation of hyphal growth of the cultivated mushroom, *Agaricus bisporus* (Lange) Sing. *Mushroom Science* **8**, 623-629.

SCHWALB, M.N. (1971). Commitment to fruiting in synchronously developing cultures of the basidiomycete *Schizophyllum commune*. *Archives of Microbiology* **79**, 102-107.

SCHWALB, M.N. (1974). Effect of adenosine 3',5'-cyclic monophosphate on the morphogenesis of fruit bodies of *Schizophyllum commune*. *Archives of Microbiology* **96**, 17-25.

SCHWALB, M.N. (1978). Regulation of fruiting. In *Genetics and Morphogenesis in the Basidiomycetes*, pp. 135-165. Edited by M.N. Schwalb & P.G. Miles. New York: Academic Press.

SHAW, W.L. & MILES, P.G. (1970). Inhibition of the development of *Schizophyllum commune* germlings by the ammonium ion. *Plant and Cell Physiology* **11**, 487-497.

SMYTHE, R. (1973). Hyphal fusions in the basidiomycete *Coprinus lagopus* sensu Buller: I. Some effects of incompatibility factors. *Heredity* **31**, 107-

111.
SUMMERBELL, R.C., CASTLE, A.J., HORGEN, P.A. & ANDERSON, J.B. (1989). Inheritance of restriction fragment length polymorphisms in *Agaricus brunnescens*. *Genetics* **123**, 293-300.
TAN, K.K. (1978). Light induced fungal development. In *The Filamentous Fungi*, **Vol. 3**, pp. 334-357. Edited by J.E. Smith & D.R. Berry. New York: Wiley.
ULLRICH, R. & ANDERSON, J.B. (1978). Sex and diploidy in *Armillaria mellea*. *Experimental Mycology* **2**, 119-129.
UNO, I. & ISHIKAWA, T. (1973). Purification and identification of the fruiting-inducing substances in *Coprinus macrorhizus*. *Journal of Bacteriology* **113**, 1240-1248.
VOLK, T.J. & LEONARD, T.J. (1989). Experimental studies of the morel. I. Heterokaryon formation between monoascosporous strains of *Morchella*. *Mycologia* **81**, 523-531.
VOORHEES, D.A. & PETERSON, J.L. (1986). Hypha- spore attractions in *Schizophyllum commune*. *Mycologia* **78**, 762-765.
WESSELS, J.G.H. (1965). Morphogenesis and biochemical processes in *Schizophyllum commune* Fr. *Wentia* **13**, 1-113.

CHAPTER 4

PRODUCTION OF A NOVEL WHITE *FLAMMULINA VELUTIPES* BY BREEDING

Yutaka Kitamoto, Masato Nakamata
and Paul Masuda

Department of Agricultural Chemistry,
Tottori University, Koyama 680, Japan.

1. INTRODUCTION

Flammulina velutipes (*Enokitake*) is the most popular bottle cultured mushroom in Japan. In 1989, the highest annual production of 83,000 tons was recorded, which had exceeded that of the fresh *Lentinus* mushroom (Ohashi, 1991).

The increase in production of *Enokitake* mushroom began in the early 1950's. Recent advances, particularly in cultivation technology, have been largely responsible for the increased mushroom yields. Automated machinery is now commonplace in mushroom plants with some of the latest facilities consistently producing 1,000 tons or more annually. For the past 35 years leading up to 1985, breeding programmes for the *Enokitake* mushroom were based on the selection of genetic variants which produced fruit bodies of desirable quality and appearance and which appeared spontaneously but rarely during commercial cultivation. The selected fruit bodies would then be used as a source material for the preparation of the next generation of seed spawn. ("Senbatsu" method) (Zennyoji, 1989). However, the limit for strain improvement using this method may have been reached. Minute changes in culture conditions often resulted in decreased yields and poorer quality mushrooms when these strains were adopted. Moreover, continual subculture during seed spawn production caused strain deviations. Thus, impetus has come from commercial source for the breeding of a new strain with favorable properties such as high stability, high productivity and high quality white-coloured fruit bodies. In 1985, we successfully produced a novel strain, M-50, which is the first breed of white fruit-body forming strains of *Flammulina velutipes* (Nakayama *et al*, 1987; Kitamoto, 1990). The breeding of M-50 and analogous white strains has had a major impact on industrial production these past seven years, during which time these new strains have been used exclusively for *Enokitake* cultivation.

In this chapter, the research and development involved in breeding the novel white strain, M-50, of *Flammulina velutipes* is described.

2. PROPERTIES OF CULTIVATED *ENOKITAKE* STRAINS IN RESPECT OF BREEDING

Flammulina velutipes is a white rot type of basidiomycete. In Japan, wild *Enokitake* forms clusters of fruit-bodies on the stumps or withered timbers of broad leaved trees from late autumn to spring. It is characterized by light brown or brown pilei with straw colored stipes. However, the cultivated mushroom shows different shapes and colors from the wild mushroom. Fruit-bodies of the cultivated mushroom are characterized by creamy white pilei and long stipes, somewhat resembling bean sprouts. For successful breeding of the *Enokitake* mushroom, it is essential to conserve the desired morphological features of the fruit-bodies during commercial cultivation as well as to understand the genetics and cultural requirements.

2.1. Physiological Requirements and Genetic Properties of the Cultivated Strains

2.1.1. Physiological factors. The optimum temperature for mycelial growth in wild and cultivated strains is about 22-24°C, but there are some exceptions among the different stains (Kitamoto, 1991a, b; Kitamoto & Suzuki, 1992). On the other hand, the optimum for fruit-body initiation is around 15°C although some wild strains produce fruit-bodies at 25°C. The temperature characteristics of the mushroom strains are one of the most important factors to consider in breeding because of the direct influence that temperature has on the physiological traits of the mycelia and fruit-bodies in this mushroom.

Lighting is another important environmental factor which controls the growth and development of *Flammulina* fruit-bodies (Plunket, 1956; Ashan, 1958). Light control is critical since low levels of light increase the number of the buds of fruit-bodies and promote pileus development, while excessive levels of light suppress stipe elongation and induce the brown coloration on pilei (Kitamoto, 1991a; Inatomi & Yamanaka, 1991). Whiteness of fruit-bodies is an important characteristic in term of market value. Therefore, improvement of genetic traits of strains to the response against light exposure is to be expected.

In the bottle cultivation of this mushroom, a mixture of sawdust and rice bran together with some additional supplements is normally used as the culture medium (Kitamoto, 1991a). The nutritional and biochemical

TABLE 1. The distribution of incompatibility factors among various wild and cultivated strains of *Flammulina velutipes**.

Strain	Tester Monokaryons				Incompatibility Factor Composition**
	A1B1	A2B2	A1B2	A2B1	
M-50	-	-	+	+	A1B2+A2B1
NAKANO	-	-	+	+	A1B2+A2B1
HATSUYUKI	-	-	+	+	A1B2+A2B1
K1	-	-	+	+	A1B2+A2B1
R2	-	-	+	+	A1B2+A2B1
M-45	-	-	+	+	A1B2+A2B1
JOHSHO	-	-	+	+	A1B2+A2B1
T-18	-	-	+	+	A1B2+A2B1
NANANO A2	-	-	+	+	A1B2+A2B1
NAKANO N1	-	-	+	+	A1B2+A2B1
NAKANO N2	-	-	+	+	A1B2+A2B1
EA	+	+	+	+	A3B3+A4B4
W1	+	+	+	+	A5B5+A6B6

* The dikaryotic incompatibility factor distribution were determined by the application of the di-mon mating methods, using the tester monokaryons from the NAKANO strain.

** Incompatibility factors are assigned temporary numbering.

characteristics of the mushroom strains are also important parameters to consider in any breeding programme.

2.1.2. Genetic factors. *Flammulina velutipes* is a tetrapolar basidiomycete forming clamps on the dikaryotic hyphal cells during growth (Takemaru, 1961). The origin of the first cultivated strain in Japan has not been confirmed, but it is widely recognized that SHINANO-1 was a predominant strain in the early years of cultivation (Zennyoji, 1989; Kinugawa, 1989). It appears that subsequent improvement of strains had been carried out by the repeated application of the "Senbatsu" method to lines derived from the progenitor strains, which included SHINANO-1, NAKANO and HATSUYUKI. Accordingly, as seen from the dikaryon-monokaryon (di-mon) mating analysis (Kimura, 1977) shown in Table 1, all of the strains cultivated hitherto have the same composition of incompatibility factors in the hyphal cells as those found in the NAKANO tester strain. On the other hand, wild strains differ from the tester strain in their incompatibility factor composition (Kitamoto, 1991a).

2.2. Culture Processes

In breeding mushrooms, it is important to assess the performance of the new strain under the environmental conditions which prevail during each of the sequential processes involved in cultivation.

As shown in Figure 1, five major culture processes are involved in the bottle cultivation of *Enokitake* mushroom (Kitamoto, 1991a). The first step is the spawn running ("Baiyo") which allows mycelial growth to continue for about 25 days in culture rooms maintained at 14-16°C. To compensate for the heat and carbon dioxide generated by fungal metabolic activity, cooling and adequate ventilation to prevent a built-up of CO_2 levels are required. It is necessary to maintain the temperature of the culture room several degrees lower than the optimum growth temperature (22-24°C) in order to offset the exothermic effect. The second air-conditioned step in the cultivation process is to initiate budding. When the spawn has grown well throughout the medium, the cultures are subjected to the "Kinkaki" treatment (scratching away and removal of surface mycelia in the culture bottle to induce budding). Cultures are then transferred to culture rooms maintained at 15°C to initiate fruit-body formation. This step is followed by the "Narashi" and "Yokusei"

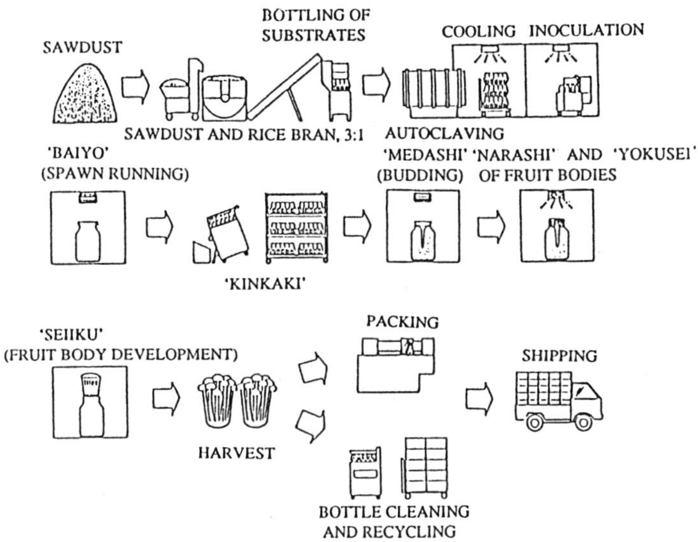

FIGURE 1. The culture processes in the commercial bottle cultivation of Enokitake mushroom.

processes for acclimating the mushroom to reduced humidity, and for decreasing the growth rate to equalize the size of fruit-bodies, respectively. These processes are carried out successively at 5°C in the same culture room. The final step is the growing up ("Seiiku") of fruit-bodies. The environment is usually maintained at 7-9°C and 75-80% relative humidity. It is probable that alterations in the temperature traits of strains in relation to each of the cultivation processes may greatly contribute towards reducing running costs and initial construction capital attributed to air-conditioning.

3. BREEDING STRATEGIES FOR NEW *ENOKITAKE* STRAINS

The marketable characteristics demanded of new strains of *Enokitake* mushroom have continually been changing from the time the mushroom was first cultivated about 35 years ago. However, the more generally desirable morphological features of fruit-bodies were roughly defined more than 10 years ago. Therefore, the inheritance into new strains of genetic traits possessed by existing strains is a fundamental part of current breeding strategy. Either the mating method or cell fusion can be adopted for the production of dikaryotic hybrids. An experiment was carried out to compare the manner of gene expression in the daughter dikaryotic stocks hybridized by mating and by cell fusion techniques using two monokaryotic stocks of *Coprinus cinereus* as a test organism (Tanaka, H., unpublished results). In comparing the esterase zymograms of the fusant and the mated hybrids, the fusant zymogram was intermediate to both of the hybrid zymograms (one hybrid from each side of the mated parental mycelia). Accordingly, the application of the cell fusion method for intraspecific hybridization appears to be tedious and inefficient compared to the mating method. Thus, we have aimed to improve the efficiency of the mating method for the breeding of new *Enokitake* strains.

3.1. Objective in Breeding a New Strain

The objective in breeding a new strain is to produce one that is best suited for commercial production. Some of the desirable properties and qualities to look for in a new strain are: (1) high productivity, (2) high quality fruit-bodies, (3) favorable culture characteristics, and (4) stability of strain properties during continuous sub-culture required for seed spawn production. Among those properties listed above, high quality fruit-bodies refers to retention of a white color during development, favorable shapes and dimensions of pilei and stipes with no adhesive aerial hyphae on the base parts of fruit-body bunches, and a prolonged shelf-life. It is also important to consider strain stability in terms of maintaining high productivity and quality in the face of fluctuating environmental conditions during cultivation.

3.2. Methodology for Breeding a New Strain

3.2.1. Use of genetic traits of monokaryotic stocks in breeding. The first step in the production of hybrid strains is single spore isolation and the preparation of monokaryotic line stocks derived from those existing strains to be used for mating. As described in the preceding section, all of the monokaryotic stocks derived from previously cultivated strains can be divided into only four mating types. Thus, the determination of incompatibility factors for monokaryotic stocks enhances the following hybridization operations. The use of micro-manipulation may greatly increase productivity and reliability during preparation of monokaryotic stocks.

All of the pure line mushroom breeds derived by mating show some different traits because each of the monokaryons belonging to the same parental line are genetic variants resulting from the meiotic process which occur during spore formation. Figure 2 shows a comparison of the esterase isozyme pattern of each of the monokaryotic mycelia derived from the spores of a fruit-body of the JOHSHO strain of *Flammulina velutipes* (Kitamoto *et al*, 1986). Small differences were observed among the zymograms of monokaryons having the same incompatibility factors. Variations in the genetic traits of the monokaryons might be enlarged among stocks with different incompatibility factors.

The germination of basidiospores is affected by their hereditary nature. Figure 3 shows an analysis of germinated spores having different incompatibility factors from fruit-bodies of the M-45 strain of *Flammulina velutipes*, in which the spores were isolated with a micro-manipulator

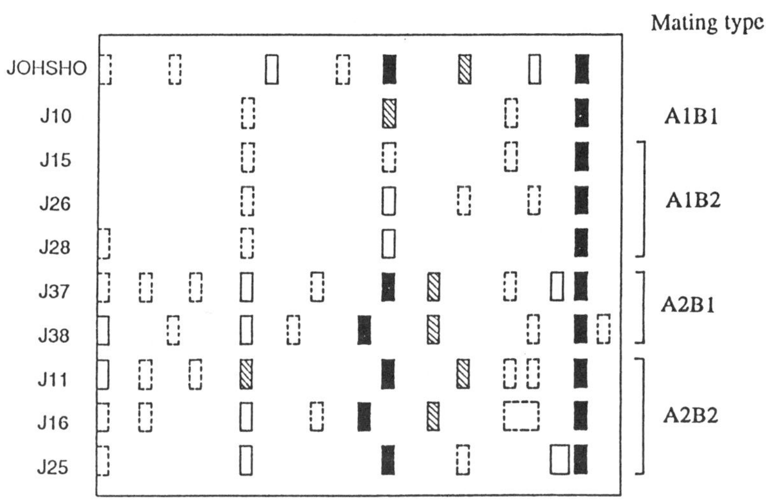

FIGURE 2. Zymograms showing relative locations of esterase bands in various monokaryotic stocks derived from a dikaryotic strain (JOHSHO) of *Flammulina velutipes.* Relative staining intensity: ■ → ▨ → □ → ⬚

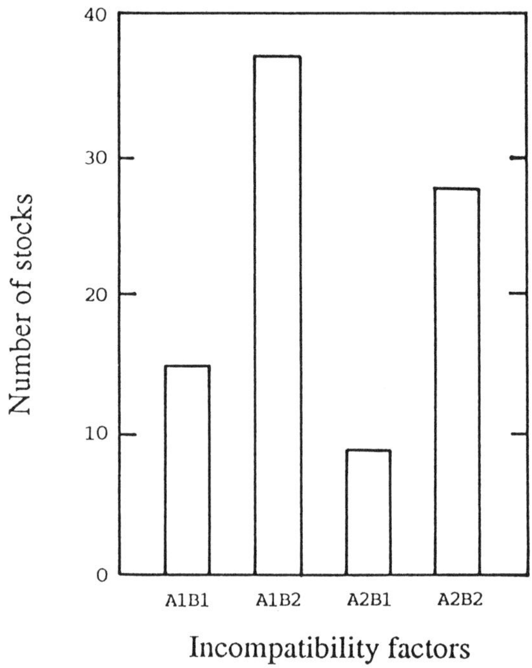

FIGURE 3. Frequency distribution of four incompatibility factor combinations among the germinated basidiospores from a fruit-body of the dikaryotic M-45 strain of *Flammulina velutipes*.
* Incompatibility factors are assigned temporary numbering.

(Kitamoto, 1991b). The rate of germination among experiments usually deviated over a range of 5-50%. Remarkable differences in the germination rate observed among the spores of the four different mating types suggests that linkage of the B incompatibility factor with genetic factor(s) controlled the germination process in the basidiospores. Due to variable germination rates among the four incompatibility factors, increased monokaryotic isolates are required among the low germinating mating types in order to maintain even representation from each of the mating groups.

The growth of monokaryotic mycelia is affected by one or more genetic factors, which might be linked with their incompatibility factors. Figure 4 shows an example for the growth characteristics of monokaryons obtained from the spores of a fruit-body of the M-45 strain of *Flammulina velutipes*

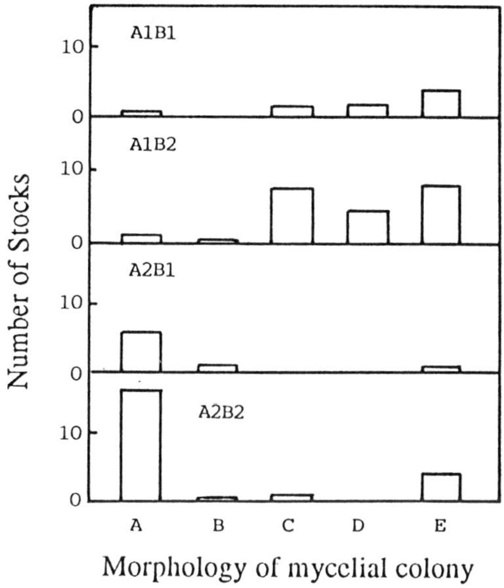

FIGURE 4. Frequency distribution showing a correlation between incompatibility factors and the morphology of mycelial colonies in various monokaryons derived from the dikaryotic M-45 strain of *Flammulina velutipes.* The colonies were grown on PDA medium.
Incompatibility factors are assigned temporary numbering.
A: Stocks growing fast to form soft colonies.
B: Stocks growing fast to form tight colonies.
C: Stocks growing slow to form faint colonies.
D: Stocks growing slow to form compact colonies.
E: Stocks growing poorly to form mycelial masses.

(Kitamoto, 1991b). Stocks of mycelia having the incompatibility A2 factor usually grew faster and formed more uniform colonies than stocks with the A1 factor.

There is an empirical rule which applies to the expression of temperature traits among hybrids of monokaryotic stocks. As shown in Table 2, the mating between two compatible monokaryotic stocks, both of which had high temperature optima, reproduces the high temperature traits among dikaryotic hybrids. However, hybridization between the high-and-low or the low-and-low combinations produced variable results. The application of pre-screening based on the genetic traits of the monokaryotic stocks may greatly increase

TABLE 2. An empirical rule showing high optimal temperature expressions among dikaryotic hybrids produced by different mating combinations of *Flammulina velutipes**.

Mating combinations	Incompatibility factors	High temperature hybrid number	High temperature hybrid rate
H x H	A1B1 + A2B2	3/3	
	A1B2 + A2B1	8/8	
		Total 11/11	100%
H x L	A1B1 + A2B2	8/10	
	A1B2 + A2B1	1/2	
		Total 9/12	75%
L x H	A1B1 + A2B2	2/5	
	A1B2 + A2B1	2/3	
		Total 4/8	50%
L x L	A1B1 + A2B2	2/4	
	A1B2 + A2B1	2/3	
		Total 4/7	57%

*Strains with optimal temperatures higher or equal to 22.5°C were designated as high temperature stocks (H). Strains with optimum temperatures below 22.5°C were considered low temperature stocks (L). The culture test was performed on PDA medium at various temperatures.

the efficiency of breeding.

3.2.2. The methods for mating. There are two classes of mating systems, "self" (=pure line) and "cross" matings, which describe hybridization within the same or between different monokaryon lines, respectively.

The effectiveness of the two mating methods was evaluated by comparing the level of deviations which occurred among the groups of hybrid stocks produced by "self" and "cross" matings, with reference to the fruit-body yields and the numbers of fruit-body buds produced in bottle cultures. As shown in Figure 5, somewhat wider variations in both of the reference features were observed among the "cross" mating group of hybrids compared with those in the "self" mating group, although both of the mother strains were progenies of the ancestral SHINANO-1 strain obtained by successive "Senbatsu"

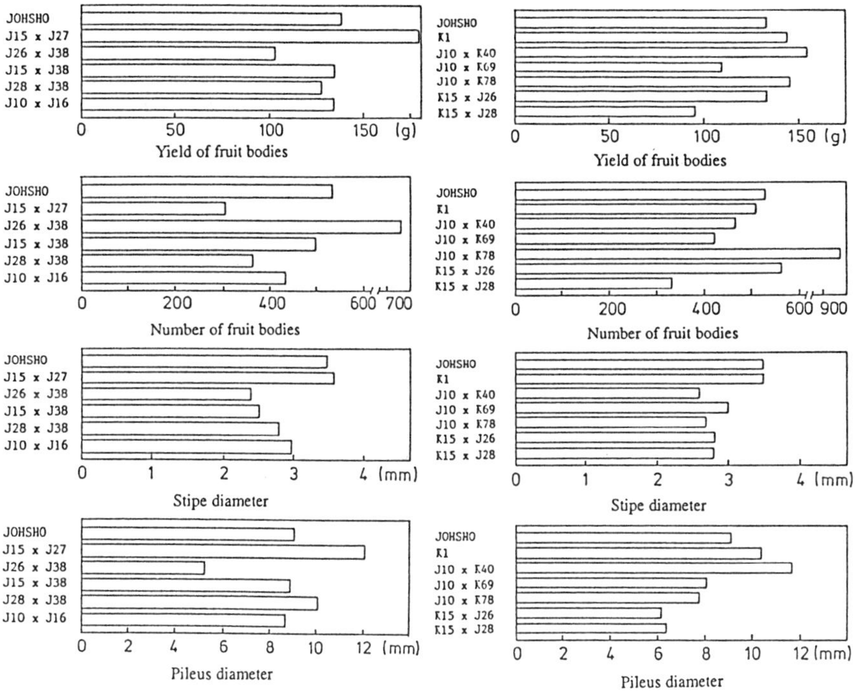

FIGURE 5. Variations in some morphological traits of hybrid stocks produced by mating between the monokaryons from one or two cultivated strains of *Flammulina velutipes*.
(a) Mating between monokaryons from JOHSHO.
(b) Mating between monokaryons from JOHSHO and K1.

selection. Figure 6 illustrates the remarkable variation in the shapes of the fruit-bodies of hybrids produced by "cross" mating involving two lines of monokaryotic stocks from the two parental dikaryotic strains, NAKANO and K1. It is reasonable to expect wide variations in morphological and culture characteristics among the "cross" mating hybrids. However, the use of "self" mating is effective for achieving minute improvements among those commercial strains under active cultivation.

Figure 7 illustrates the distribution of growth temperature variants among hybrid stocks produced by "self" mating between monokaryotic stocks derived from the M-50 strain. Wide and stable variations in the growth

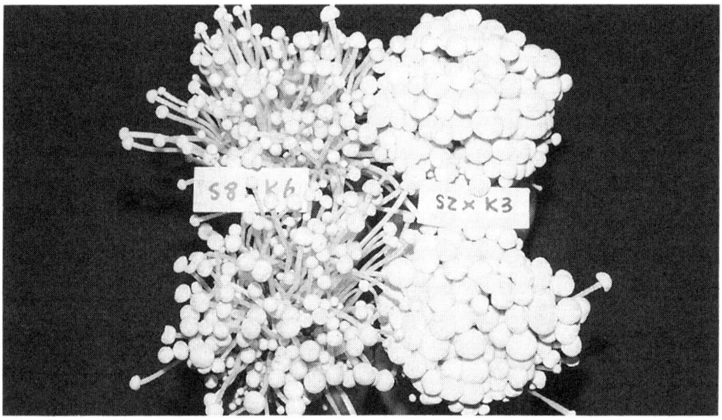

FIGURE 6. Various morphological traits of fruit-bodies in the hybrid stocks produced by mating between monokaryons from the dikaryotic strains of NAKANO(S) and K1(K).

to subject the faster growing dikaryotic mycelium to the screening process for selecting new breeding stocks.

3.3. Breeding of the White Strain

3.3.1. Discovery of white fruit-body forming hybrids. When grown in the light, the strains of *Flammulina velutipes* hitherto cultivated in Japan produce brown fruit-bodies similar to those of the wild mushroom (Figure 8). However, market needs have pressed for the development of new strains and improved cultivation methods to produce mushrooms with fruit-bodies which are almost white in color. Color development is due to the accumulation of phenolic pigments in fruit-bodies. When the mushroom is cultivated in the light, increased activity of phenol oxidase is observed (Nakayama, I., unpublished data). As described in section 3.3.3., there is a positive correlation between the activity of phenol oxidase and the number and/or the yield of fruit-bodies (Kitamoto, 1991a). Cultivation of the mushroom in dim light may reduce the color development in fruit-bodies. However, the brown coloration in the basal part of stipes is inevitable in colored strains (Figure 9).

In our preliminary experiments to survey color development in cultivated strains, it was estimated that a reduction in the color tone of fruit-bodies had

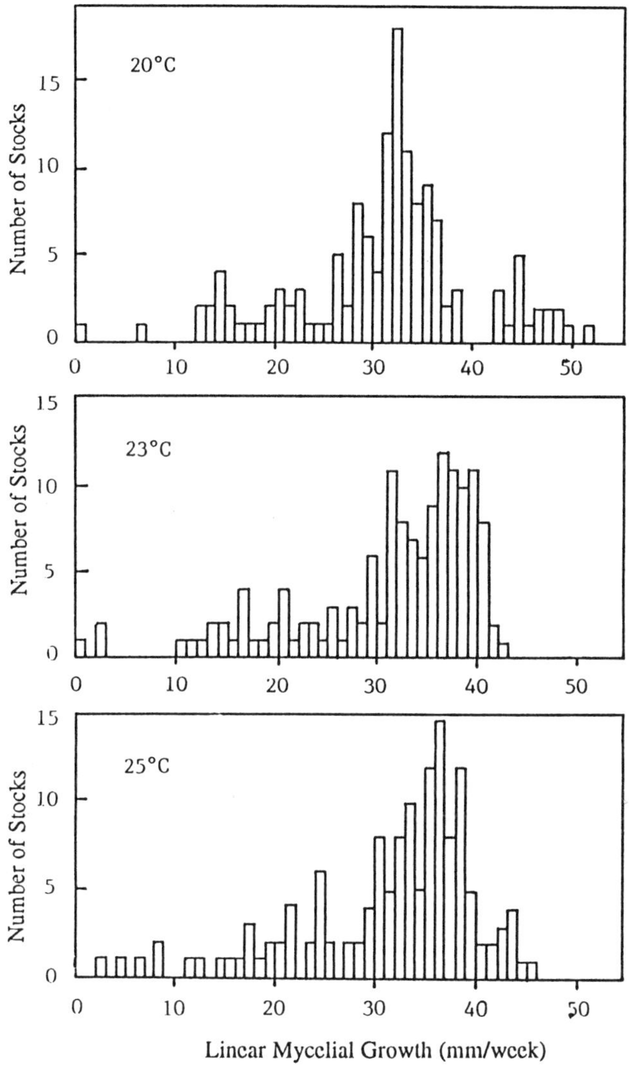

FIGURE 7. Frequency distribution showing the growth temperature variants of hybrid stocks produced by self mating between the monokaryotic stocks derived from M-50 strain. The number of stocks showing various mycelial growth rates (one week on PDA medium) were grouped into 1mm/week intervals.

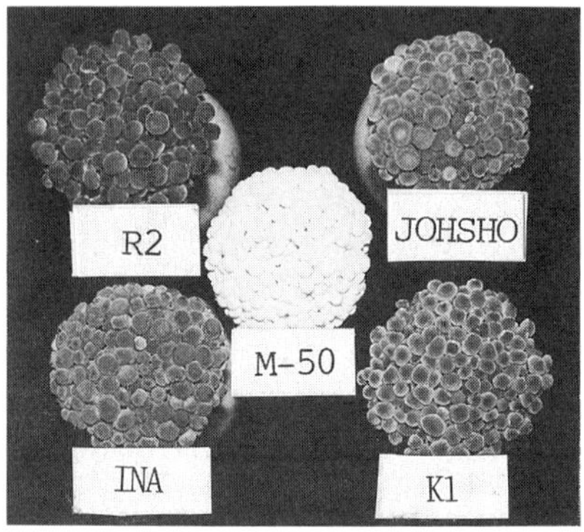

FIGURE 8. Color of fruit-bodies of various strains grown under 300 lux of light illumination.

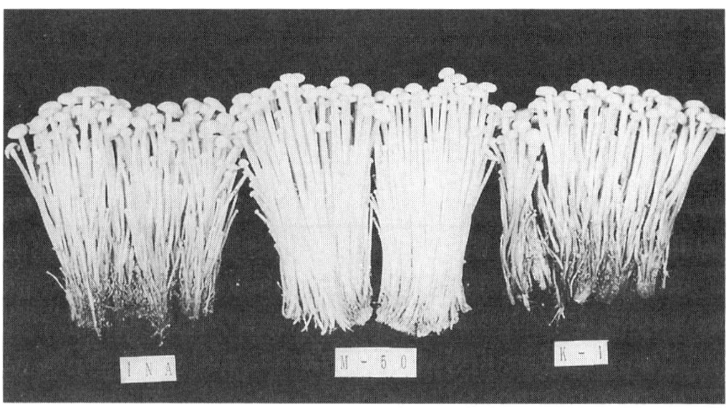

FIGURE 9. Brown coloration in the basal parts of fruit-body bunches in various strains grown under dim light conditions (>10 lux).

occurred in more recent generations. This survey also suggested the possibility of further improvement in the "whiteness" of fruit-bodies. Thus, we carried out a small scale cultivation test under light illumination using over one hundred of the hybrid stocks which had been produced by "self" and "cross" matings between the monokaryotic stocks of several cultivated strains (Nakayama *et al*, 1987; Kitamoto, 1990). The results are summarized in Table 3.

TABLE 3. Morphological and physiological characteristics of various hybrid dikaryons of *Flammulina velutipes* produced by self and cross mating of monokaryotic mycelia derived from wild and commercial strains.

Strains or Hybrids	No. of Stocks	Fruit-body Yields ≥140g/bottle	Number of fruit-bodies ≥500/bottle	Stipe Diameter ≥3 mm	Pileus Diameter ≥9 mm	White Fruit-body Formation
Commercial Strains*	8	2	4	4	4	0
S x S	2	0	1	0	2	0
K x W1	4	1	0	2	0	0
S x W2	6	1	2	1	3	0
J x J	43	9	7	9	8	6
J x K	18	5	5	3	9	7
J x R	16	1	1	1	2	1
R x K	7	0	0	0	1	0
Total	104	19	20	20	29	14

* The commercial strains tested were NAKANO(S), HATSUYUKI, K1(K), R2(R), JOHSHO(J), T-18 and NAKANO A2. W1 and W2 are wild strains.

Eight of the cultivated strains including NAKANO(S), K1(K), R2(R) and JOHSHO(J) formed only brown fruit-bodies like those of the wild strains, W1 and EA. The "self" mated hybrids derived from the monokaryons of the strains above, except for JOHSHO, produced brown fruit-bodies. The "cross" mated hybrids derived from the monokaryotic crosses K x W1, S x EA and K x R also formed brown fruit-bodies. However, the formation of white fruit-bodies was unexpectedly found among some hybrid stocks formed by the mating of J x J, J x K and J x R. The highest rate of appearance of the white stocks was observed among "cross" mating hybrids of J x K. This group of hybrids may also have additional advantages over white hybrids from other crosses in terms of yield, and the number and shape of fruit-bodies.

3.3.2. Screening for white strains. The selection of white fruit-body-forming stocks from among the test hybrids was easily achieved by employing test cultivations in the light as described above. It is possible to shorten the test period for the selection of white stocks to less than 40 days by using small volume culture tubes in place of the standard plastic culture bottles adopted for commercial cultivation.

3.3.3. Screening for high yield strains. Productivity is one of the most important considerations of any breeding programme and must be taken into account along with other desirable traits. However, introduction of standard cultivation tests to select high yielding strains from a large number of test hybrids requires the installation of a large-scale pilot plant. Thus, in order to increase the overall efficiency of the screening process, the application of pre-screening methods to vegetative mycelium was examined (Kitamoto, 1991a).

It is reasonable to expect a correlation between mycelial growth of a strain in a defined culture medium and the resulting yield of fruit-bodies since the growth of fruit-bodies may well depend on the supply of nutritional substances which have accumulated in the vegetative mycelia prior to the initiation of fruiting (Kitamoto and Suzuki, 1992). The mass of vegetative mycelia is likely to reflect the capacity for depositing storage materials for later use as growth substrates for the fruit-bodies. Figure 10(a) shows a typical correlation between the rate of linear mycelial growth in test tube cultures and fruit-body yields in standard bottle cultures using sawdust medium of the same composition as that employed for *Enokitake* cultivation. The correlation factor from the experiment (r=0.625) was not enough for choosing high yield strains with any great degree of certainty. Nevertheless, if high yield strains are designated on the basis of a correlation between mycelial growth rate and yield above a threshold yield level of 130g/bottle, this approach may still be useful for condensing the number of potentially satisfactory hybrid stocks in the remaining population. However, no apparent correlation was found between linear mycelial growth on agar medium or mycelial dry weight in liquid culture and the fruit-body yield on sawdust medium in bottle cultures.

Levels of phenol oxidase activity in the vegetative mycelia of test stocks may also be useful as a criterion for pre-screening high yield strains (Kitamoto, 1991a). Figure 10(b) shows the correlation between the two variables in hybrids of "cross" mated monokaryons of the J and K lines of *Flammulina velutipes*. When enzyme activities in mycelia grown on liquid culture were plotted against the fruit-body yields in bottle cultivations, a correlation factor of 0.58 was obtained.

The application of more than two criteria for pre-screening of high yield strains might increase the reliability of the screening procedure.

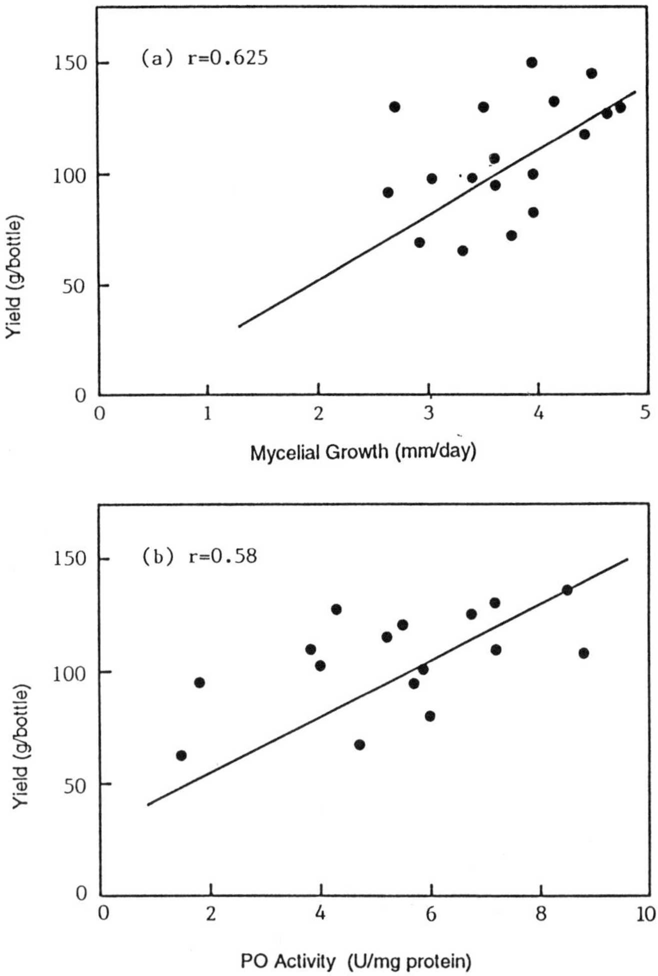

FIGURE 10. Mycelial growth and phenol oxidase activity as criteria for screening high productivity strains among a number of dikaryotic hybrids of *Flammulina velutipes*.

(a) Correlation between mycelial growth of various hybrids on sawdust medium and fruit-body yields in bottle cultivation.

(b) Correlation between phenol oxidase activity in vegetative mycelia of various hybrids grown in liquid culture and fruit-body yields in bottle cultivation.

4. THE WHITE STRAIN, M50: GENETIC AND CULTURE CHARACTERISTICS

Figure 11 shows the fruit-bodies of a novel white strain, M-50, which has been produced by applying mating and screening techniques described in this work. This novel strain was a hybrid obtained by "cross" mating between two different monokaryons, J26 and K15, and is the first officially registered white strain of *Enokitake* mushroom (Nakayama *et al.*, 1987; Kitamoto, 1990; 1991a). The morphological traits of M-50 fruit-bodies resemble the parental JOHSHO strain, in possessing a round pileus and a white stipe of about 3 mm diameter. The basal parts of fruit-body bunches are also white-colored, with little evidence of adhesive aerial hyphae. The mushroom has a prolonged fresh shelf-life of over three weeks under conditions of cold storage. The vegetative mycelium of this strain also exhibits higher phenol oxidase activity than mycelia of colored cultivated strains, and has a high capacity for fruit-body formation. Furthermore, the optimum growth temperature of this strain is 1-2°C lower than that of the colored strains. Discovery of the white strain has allowed the introduction of the light controlled cultivation (Kitamoto, 1991a), which could not be applied to colored stocks.

The color characteristic of the M-50 white strain is hereditary since all of the pure line progenies of this strain formed only white fruit-bodies.

FIGURE 11. Fruit-bodies of the white strain (M-50) in bottle cultivation.

FIGURE 12. Appearance of brown color fruit-bodies from the population of bottle cultures of a white hybrid (FY20 x R40). All culture bottles were inoculated using seed spawn which was sub-cultured several times following hybridization.

However, the white fruit-body trait is not a stable feature for all of the white hybrids. In the case of the hybrid, FY20 x R40, the white fruit-body-forming capability was lost in parts of the daughter mycelium after several subcultures following hybridization (Shiratori, R., unpublished data). Eventually, white and brown fruit-bodies occasionally appear together in the same culture bottle inoculated with sub-cultured spawn (Figure 12).

5. MECHANISM OF WHITE FRUIT-BODY FORMATION

The M-50 strain forms light brown pin head-type fruit-bodies when cultivated in the light, demonstrating that the strain is not an albino mutant.

Young fruit-bodies showed the same or higher levels of phenol oxidase activity compared with those of the colored strains and become white during development. During the course of biochemical testing, the fruit-bodies of white strains showed higher staining activities against triphenyl tetrazolium chloride (TTC) compared to those of colored strains. This could indicate the existence of a high activity color reduction system capable of coupling with

TABLE 4. Distribution of superoxide dismutase among different strains of *Flammulina velutipes*.

Strain	Color of Fruit-bodies*	Specific activity (U/g dry cells)
M-50	white	3203
M-50 self mating	white	815-14910
25	brown	639
26	brown	1250
NAKANO	brown	929
R2	brown	1900
HATSUYUKI	brown	1040

* Fruit-bodies were cultivated under light illumination.

the TTC reagent in the cells of white fruit-bodies. Subsequently, as shown in Table 4, it was found that M-50 and daughter white dikaryons produced by "self" mating normally exhibited higher superoxide dismutase activity (SOD) compared with colored strains (Kitamoto, 1990; 1991a). We have assumed the existence of a mechanism which is capable of suppressing the oxidation of phenol compounds by oxygen radicals with high level of SOD activities in the white fruit-bodies. The SOD activity was increased several times by the light as compared to that of the dark grown mycelia of the white strain.

6. CONCLUSION

For the past 35 years, leading up to 1985, the breeding of the *Enokitake* mushroom had been carried out solely through the application of the "Senbatsu" method. This is a rudimentary way of obtaining suitable genetic variants by selecting fruit-bodies of good appearance from among all those present in the cultivation and using them to prepare the next generation of seed spawns. This chapter describes the adoption of mating techniques to produce the first white fruit-body strain (M-50) of *F. velutipes*. This was achieved by initially preparing large numbers of hybrids, and then using cultivation tests to select those hybrids with desirable traits. Thus, the final product is less the result of a logical scientific approach than a course of trial and error. A similar situation may occur in the breeding of the *Lentinus* mushroom.

To break away from the level of empiricism which prevails in current breeding programmes, it is necessary to improve our knowledge of genetic expression in heterokaryotic mushroom mycelia. It is also important to understand why genetic deviation in dikaryons is occasionally induced by sub-culture of vegetative mycelia. A practical method for selecting strains with desirable traits from large numbers of test hybrids is also an essential requirement for successful mushroom breeding programmes.

REFERENCES

ASHAN, K. (1958). The production of fruit bodies in *Collybia velutipes* II. Further studies on the influence of different culture conditions. *Physiologia Plantarum* **11**, 312-328.

INATOMI, S. & YAMANAKA, K. (1991). Effect of light on the cultivation at different processes of fruit-body formation in *Flammulina velutipes*, a colored strain "NAKANO". *Abstracts of the 6th Meeting of the Society for Mushroom Technology of Japan* p. 31. Tokyo, Japan.

KIMURA, K. (1977). Buller's phenomenon in mushrooms. *Iden* **31**, 29-34.

KINUGAWA, K. (1989). "*Flammulina velutipes*". In *Collected Data of Plant Genetic Resources*. pp. 965-967. Edited by K. Matsuo. Tokyo: Kodansha.

KITAMOTO, K. (1990). Effect of light on fruit-body development as a basis of fungal cultivation. *Abstracts of the IUMS Congress:* p.69. *Bacteriology & Mycology.* Osaka, Japan.

KITAMOTO, Y. (1991a). "*Flammulina velutipes* - Its Biotechnology". In *Fundamental Sciences and The Latest Biotechnology of Mushrooms*. pp. 223-231. Edited by the Editorial Committee of the Society for Mushroom Technology of Japan. Tokyo: Noson-Bunkasha.

KITAMOTO, Y. (1991b). "Physiology of Mushrooms". In *Year Book for Mushrooms 1991*. pp. 71-78. Edited by H. Ohashi. Tokyo: Noson-Bunkasha.

KITAMOTO, Y., NAKAYAMA, I., KAWASAKI, E., NAKAMATA, M. & ICHIKAWA, Y. (1986). Variations in morphological and biochemical traits of hybrids by mating between different lines of monokaryons in

Flammulina velutipes. *Abstracts of the Annual Meeting of the Agricultural Society of Japan*. p. 39. Tokyo, Japan.

KITAMOTO, Y. & SUZUKI, A. (1992). "*III. Physiology*". In *Kinokogaku (Science of Mushrooms)*. Edited by H. Furukawa . Tokyo: Kyoritsu-Shuppan. *In press*.

NAKAYAMA, I., SHIMADA, S., NAKAMATA, M. & KITAMOTO, Y. (1987). Production of a novel strain of *Flammulina velutipes* by mating. *Abstracts of the Annual Meeting of Society of Agricultural Chemistry of Japan*. p. 612. Tokyo, Japan.

OHASHI, H. (1991). The Statistics of Annual Mushroom Production in Japan. In *Year Book of Mushrooms 1991*. pp. 402-426. Edited by H. Ohashi. Tokyo: Noson-Bunkasha.

PLUNKET, B.E. (1956). The influence of factors of the aeration complex and light upon fruit body form in pure cultures of an agaric and polypore. *Annals of Botany* **20**, 563-586.

TAKEMARU, T. (1961). Genetic studies on fungi, X. The mating system in Hymenomycetes and its genetic mechanism. *Biological Journal of Okayama University* **5**, 227-273.

ZENNYOJI, A. (1989). Breeding of the varieties of Japanese mushrooms, especially, *Lentinus edodes*, for production in house and conservation of the cultures. *Recent Advances in Breeding* **30**, 64-73.

CHAPTER 5

PHYSIOLOGY AND THE BREEDING OF *FLAMMULINA VELUTIPES*

Kenjiro Kinugawa

Department of Agronomy, Kinki University,
Nakamachi, Nara-shi, Nara 631, Japan.

1. INTRODUCTION

The basidiomycete, *Flammulina velutipes* (Curt. ex Fr.) Sing., is one of the most popular edible mushrooms in Japan. In nature, this white-rot fungus grows on dead trunks and stumps of broad-leaved trees and, more rarely, on dead stumps of conifers. In temperate zone countries, the mushroom is collected as a food source by rural people from late autumn to spring. The fungus is cultivated in Japan using highly improved spawn, and produces bunches of long, white fruit bodies with tiny umbrella- to bell-shaped tops (Figure 1). Production continues all year round in air-conditioned mushroom houses. Total production throughout Japan in 1991 exceeded 500 billion yen (US$3.8 billion) in value.

The main body of this article is based on experimental results accrued in the author's laboratory, most of which are consistent with data reported by several other authors. However, for the sake of brevity, comparison with data reported elsewhere has been kept to a minimum.

FIGURE 1. Commercial stock of *Flammulina velutipes*, Nakano-JA.

2. PHYSIOLOGY OF MYCELIAL GROWTH AND FRUITING

2.1. pH

The pH of several media changed after autoclaving and, in the case of media originally adjusted to pH values above ca. 4, also shifted to more acidic values with time of culture. Mycelial growth occurred within a range from ca. 3 to ca. 11 when tested on several different media, although mycelia remained viable even below pH 3. Maximum growth was observed between pH 6-7. The maximum yield of fruit-bodies occurred at pH ca. 6 both in stationary cultures and in cultures where spent medium was replaced by fresh material.

2.2. Vitamins

Although many fungi require vitamins such as thiamine, biotin, and those of the B_6 complex (pyridoxal, pyridoxal phosphate, pyridoxine), *F. velutipes* is semi-heterotrophic for thiamine and autotrophic for others. Mycelia were able to grow on a basal medium (BM) consisting of $MgSO_4.7H_2O$, 15 mg; K_2HPO_4, 30 mg; $ZnSO_4.7H_2O$, 0.2 mg; $(NH_4)_6MoCr_4.7H_2O$, 0.2 mg;

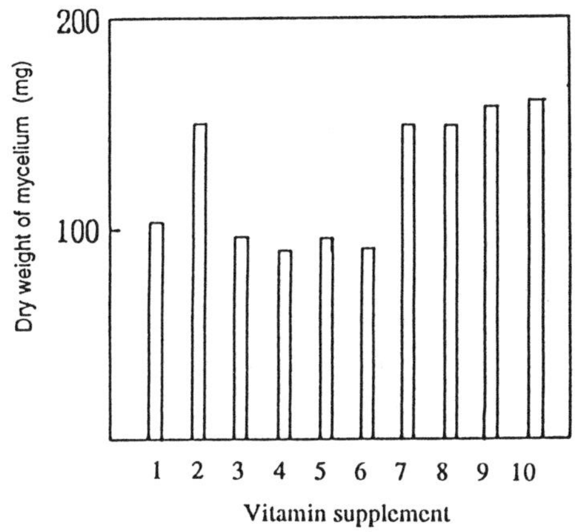

FIGURE 2. Effect of various vitamins on the mycelial growth of *F. velutipes*. Supplement: 1. none; 2. Vit B_1; 3. folic acid; 4. Vit. B_{12}; 5. Vit K_3; 6. Vit E; 7. Vit B_1+folic acid; 8. Vit B_1+Vit B_{12}; 9. Vit B_1+Vit K_3; 10. Vit B_1+Vit E.

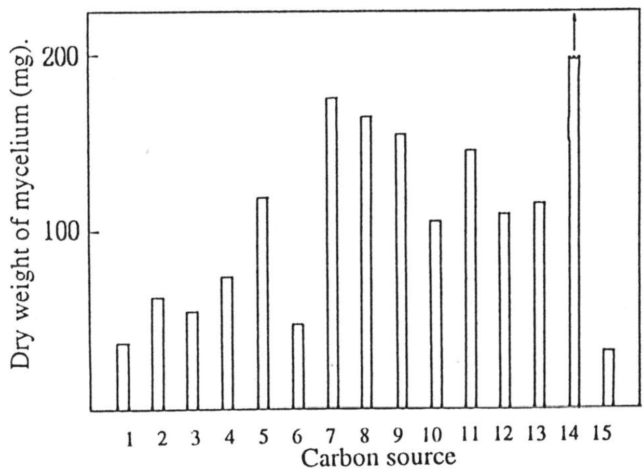

FIGURE 3. Effect of various carbon sources on the mycelial growth of *F. velutipes*. 1. none; 2. ribose; 3. arabinose; 4. xylose; 5. mannose; 6. galactose; 7. glucose; 8. fructose; 9. mannitol; 10. maltose; 11. sucrose; 12. lactose; 13. raffinose; 14. starch; 15. inulin.

$CaCl_2.2H_2O$, 5 mg; $MnSO_4.4H_2O$, 0.2 mg; distilled water, 100 ml, supplemented with sucrose 2 g and Na-glutamate 0.1 g. Growth was markedly improved by the addition of 40 μg thiamine before or after autoclaving the medium (Figure 2). Addition of any one of the vitamins riboflavin, folic acid, niacin, B_{12}, C, E, or K_3 to the medium did not promote mycelial growth. Furthermore, no synergistic effects were seen when several combinations of vitamins were tested. Apparent stimulation of fruiting was also observed on media supplemented with thiamine. Addition of thiazole and pyrimidine, each of which is a moiety of the thiamine molecule, was also effective in enhancing mycelial growth and fruiting.

2.3. Carbon Sources

Organic carbon sources are essential for all living things which lack chlorophyll. Although *F. velutipes* can utilize the carbon component of many organic compounds as a source of carbon, different organic compounds are utilized with widely varying efficiencies. However, BM medium supplemented with Na-glutamate and thiamine supported mycelial growth to some degree since the fungus can use the carbon skeleton of glutamic acid as carbon source. When BM medium containing glutamate and thiamine was supplemented with various additional carbon sources, the highest growth rate was observed with soluble starch while glucose, fructose, mannitol and sucrose supported vigorous fungal growth. Mannose, maltose, lactose and raffinose enhanced fungal growth to a considerable extent, ribose, arabinose and xylose to a lesser extent, and galactose and inulin not at all (Figure 3). Well developed colonies often fruited earlier and gave better yields of fruit bodies. The final biomass yield depended upon the amount of carbon nutrient available in the medium. Addition of fresh medium containing sucrose or glucose to carbon-depleted cultures rejuvenated the colony but postponed fruiting. This is compared to glucose repression.

2.4. Nitrogen Sources

Fungal mycelium was unable to grow on medium lacking in nitrogen, even when the medium contained other essential nutrients such as carbohydrates, minerals, and vitamins in sufficient amount. Nitrogen-containing organic compounds varied considerably in their ability to support mycelial growth and initiate fruiting (Figure 4). Three groups were identified: glutamate, aspartic acid and valine were most effective in supporting growth and

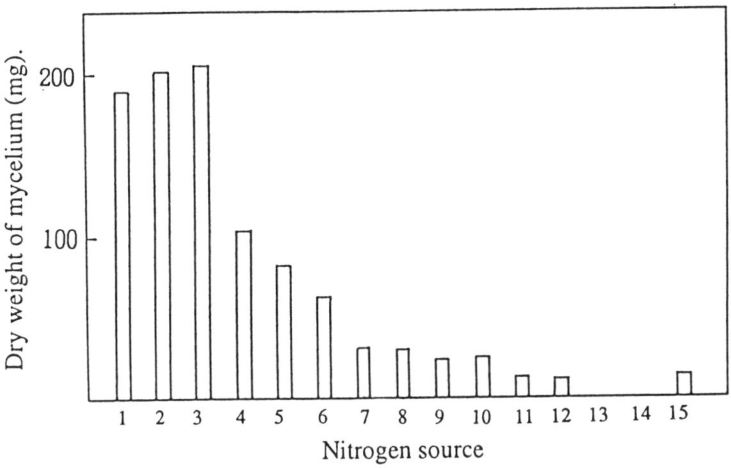

FIGURE 4. Effect of various nitrogen sources on the mycelial growth of *F. velutipes*. 1. sodium glutamate; 2. aspartic acid; 3. valine; 4. urea; 5. alanine; 6. arginine; 7. methionine; 8. tryptophane; 9. serine; 10. leucine; 11. $(NH_4)_2SO_4$; 12. NH_4NO_3; 13. $NaNO_3$; 14. KNO_3; 15. none.

initiating fruiting, urea, alanine, and arginine were effective but less so, while methionine, tyrosine, serine, and leucine supported only poor growth and fruiting levels. Addition of nitrate salts suppressed fungal growth. Peptone, which contains various amino acids and low molecular polypeptides, and is contaminated with minerals and vitamins in trace amounts, supported mycelial growth even when supplied alone. Fruiting also occurred on the well-developed colonies. Fungal mycelium grew well in a medium containing sucrose and peptone without additional thiamine supplementation. Fungal growth caused a shift to a higher pH in peptone rich media and to a lower pH in sucrose rich media. For fruiting, the upper limit of peptone concentration was 0.8g/100 ml. Under these culture conditions, the increase in mycelial yield with increasing peptone concentration followed a Liebig's exponential curve. The asymptote was dependent upon the amount of sucrose added. In contrast, in the chemically defined media where peptone was replaced by Na-glutamate, mycelial growth attained extension rate and dry weight maxima at 0.2 to 0.4 mg Na-glutamate/100 ml, then decreased rapidly at higher concentrations. When fresh liquid medium containing 2g/100ml sucrose and various amounts of Na-glutamate replaced spent medium in aged cultures,

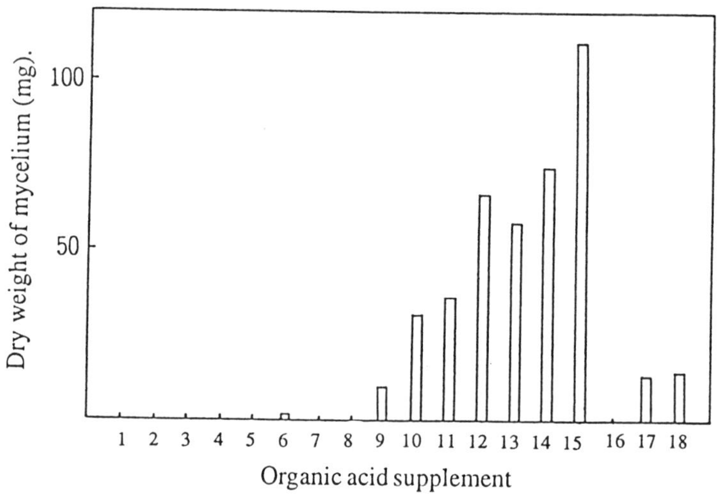

FIGURE 5. Effect of organic acids on the mycelial growth of *F. velutipes*.
1. tartrate; 2. citrate; 3. acetate; 4. formate; 5. succinate; 6. malate; 7. glycolate; 8. oxalate; 9. glucose; 10. glucose + tartrate; 11. glucose + citrate; 12. glucose + acetate; 13. glucose + formate; 14. glucose + succinate; 15. glucose + malate; 16. glucose + glycolate; 17. glucose + oxalate; 18. glucose + ATP.

maximum yields of mycelia and fruit bodies were obtained at 0.1g/100ml and 0.4mg/100ml Na-glutamate, respectively. Another experiment showed that cultures appeared to require more nutrient nitrogen at the time of fruiting than during vegetative growth.

2.5. Stimulation of Mycelial Growth by Organic Acids

Utilization of nutrients by fungi is normally more effective when they are given as a mixture than as a single compound. Thus, synergistic effects between two or more nutritive compounds are commonly observed. For example, various organic acids stimulate mycelial growth when added in trace quantities to media containing appropriate amounts of carbon and nitrogen sources. Of several organic acids added separately (0.1 mg) to 100 ml BMGGB medium (BM medium supplemented with 0.2g Na-glutamate, 2g glucose, and 40μg thiamine), malic acid was the most effective (Figure 5).

Tartrate, citrate, acetate, formate and succinate also enhanced growth while oxalate afforded no stimulation. Glycolic acid was toxic. The stimulatory effect of organic acids was also observed when the glucose in the medium was replaced by maltose, fructose or mannose. However, no enhancement of mycelial growth was observed in complex media containing undefined ingredients, probably because the organic acids were already present in adequate amounts. Fruiting also occurred earlier, and higher yields were obtained, when cultures were supplemented with organic acids.

2.6. Mineral Elements

From a fungal nutritional standpoint, mineral elements can be divided into two classes: macronutrients, which are required at concentrations ca. 10^{-3} M, and micronutrients where concentrations of 10^{-6}M or less are the norm (Wood & Fermor, 1985). Experiments to determine the mineral requirements of *F. velutipes* were subject to inaccuracies due to impurities derived from chemicals or glassware. However, fungal mycelia failed to grow on a medium lacking potassium, phosphorus, magnesium or sulphur. These macronutrients are essential to the fungus. The other mineral elements, iron, zinc, and manganese at 10^{-6} to 10^{-5} M, copper, cobalt and molybdenum at 10^{-7} M, and calcium at 10^{-4} M, were also examined. These elements were added separately or together to a basal medium consisting of 2g sucrose, 0.2g aspartic acid, 14mg $MgSO_4.7H_2O$, 30 mg K_2HPO_4, 40 µg thiamine and 100 ml distilled water. Most of the elements showed no promotional or inhibitory effects on mycelial growth. By contrast, supplementation with zinc and the presence of zinc and calcium together in the medium (Figure 6) resulted in marked stimulation of mycelial growth. In replacement experiments (where exhausted medium was replaced by fresh) resting mycelium were rejuvenated when the fresh medium contained C- and N-sources, thiamine, and a mineral supplementation containing at least P and K. No significant rejuvenation occurred when these factors were supplied separately.

2.7. Availability of urea

In BM medium supplemented separately with starch, sucrose or glucose, urea was a less effective nitrogen source compared with Na-glutamate (Figure 7). However, marked stimulation of growth occurred when urea was added to the

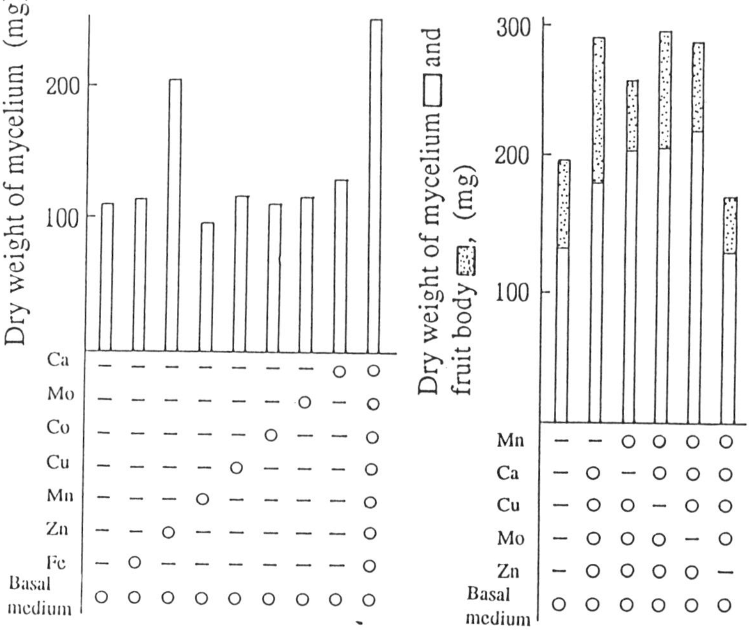

FIGURE 6. Effect of mineral elements on mycelial and fruit-body growth of *F. velutipes*.

medium together with starch and sucrose, or with starch and glucose. Growth stimulation was also observed when Na-glutamate replaced urea although to a lesser degree.

2.8. Temperature

Mycelium of certain stock cultures exhibited optimal mycelial growth at 24-25°C, very slow growth below 5°C, and no grow above 35-40°C on the medium composed of sucrose and dried yeast powder ("Ebios"). Although fruiting occurred in stationary cultures incubated below 25°C, yields were enhanced when the cultures were transferred to lower temperatures. This effect proved to be due not to the difference between the temperature levels, but to the level of the low temperature itself (10-15°C). When a well-developed colony was exposed to low temperature (10-15°C), exposure for

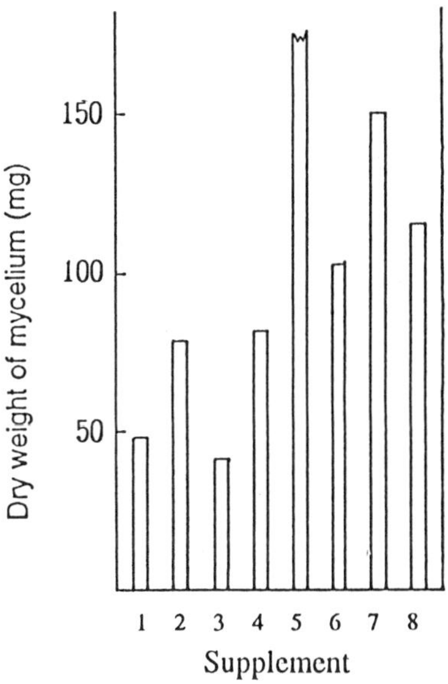

FIGURE 7. Effect of urea supplementation on mycelial growth of *F. velutipes*. Supplementation: 1. urea + starch; 2. sodium glutamate + starch; 3. urea + sucrose; 4. urea + glucose; 5. urea + starch + sucrose; 6. sodium glutamate + starch + sucrose; 7. urea + starch + glucose; 8. sodium glutamate + starch + glucose.

more than 12 hours caused a change in mycelium physiology from vegetative to reproductive and induced fruit-body formation (Kinugawa & Furukawa, 1965).

2.9. Effects of Excess CO_2 and Limited O_2

F. velutipes, like other organisms, absorbs O_2 and releases CO_2 through respiration and various other metabolic pathways. Thus, O_2 in the gaseous phase is replaced by CO_2. Diminution of ambient O_2 concentration inhibited mycelial growth and eventually stopped fruiting. High CO_2 concentrations (10-20%) have been reported to stimulate mycelial growth of some wood rotting basidiomycetes (Hintikka & Korhonen 1970: Zadrazil, 1978). A similar phenomena were also observed in *F. velutipes*. In commercial

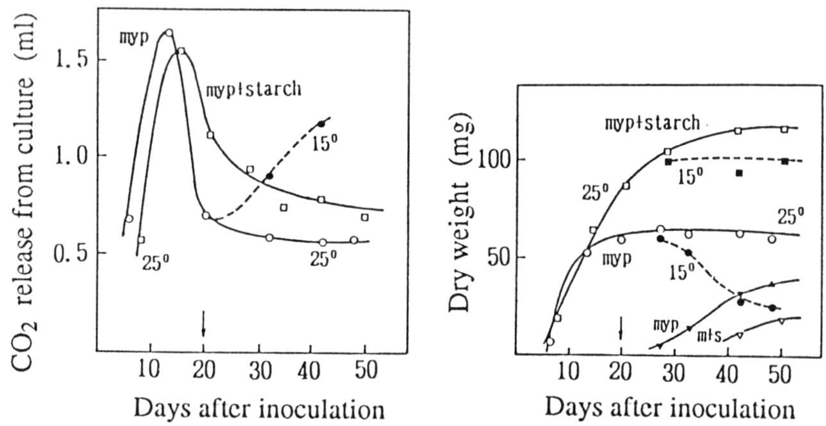

FIGURE 8. Changes in the rate of CO_2 release, mycelial yield and fruit-body yield during growth of *F. velutipes* on MYP medium and MYP medium + starch. △ = fruit-body yield.

cultivation using SR medium (115 g sawdust [*Cryptomeria japonica*], 63 g rice bran, and 110-115 ml water), exposure of cultures to high CO_2 concentration (6000 ppm for 7 days) during spawn running promoted mycelial growth and greatly decreased guttation. Exposure during fruiting decreased primordial formation and increased stipe length (Kinugawa *et al.* 1986). Half the total volume of the SR medium was space filled with water and air. When the CO_2 in the air increased up to 10-20%, the mycelial growth depended upon the level of oxygen transfer from the medium surface. These results were consistent with those reported by Plunkett (1956), Niederpluem (1963), Long (1966) and Zadrazil (1978).

2.10. Effects of Light on Fruiting

Light is known to be an important factor affecting fruiting and fruit-body morphogenesis in most higher basidiomycetes (reviewed by Eger-Hummel, 1980). For *F. velutipes*, however, light was not essential to induce fruit body primordia, but was effective in increasing the number of mature fruit bodies except in cases where the culture was poorly aerated (Kinugawa, 1977). Light was also effective in shortening the stipe length, promoting browning of fruit-bodies, and in causing early opening of caps. Opening or removal of the cap resulted in a sudden slowing and eventual cessation of stipe growth (Gruen 1976, 1979). Since white and longer fruit-bodies meet with public favour in the markets of Japan, lighting in commercial cultivation operations has been reduced to a minimum. In the past decade, white commercial varieties have been developed which do not undergo browning when exposed to light (e.g. the variety Nakano-JA).

3. PHYSIOLOGICAL PROCESSES DURING CULTIVATION

Mycelial dry weight increased exponentially in cultures of *F. velutipes* grown on MYP medium (malt extract, 7 g; soytone, 1g; and yeast extract, 0.5 g in 1,000 ml distilled water) (Bandoni, 1972). Assuming that the rate of CO_2 release from the mycelium served as an indicator of physiological activity, activity first increased and later decreased (Figure 8) in response to the consumption of nutrients available in the medium (Kinugawa & Tanesaka, 1990). From the beginning of cultivation, the growing mycelium actively secreted amylase in order to convert the starch in the medium into glucose. Thus, mycelium biomass reached maximum when the rate of CO_2 release dropped to a minimum at the time when absorption of nutrients by the mycelium had almost ceased. At this time, low temperature stimulation of fruiting was most effective. Addition of more carbon source to the culture postponed the reduction in CO_2 evolution and also the onset of fruiting. Growth of the fruit bodies proceeded mainly at the expense of mycelial biomass, and the respiratory activity of the mycelium revived as a result of further absorption of nutrients remaining in the medium. Just prior to fruiting, a transient increase in the nitrogen content of the medium was detected. This may have been caused by secretion of nitrogen-containing compounds from the hyphae as part of the cell components were mobilized for fruit body growth. In SR medium, amylase activity in cultures was also high from the beginning, and reducing sugar concentration followed a pattern which was virtually identical to that of CO_2 release. Activity of laccase and cellulase appeared after spawn running had proceeded to some degree. Lignin peroxidase activity was detected around the time of harvest. The use of wood (sawdust) as a nutrient source appears impractical given the current state of cultivation technology. The fungus could not utilize a beech wood specimen within 80 days following inoculation (Tanesaka, personal communication). The

TABLE 1. Outline of the cultivation system of F.velutipes in Japan.

Varieties		Spawn run[1]	Spawn removal[2]	Fruiting induction	Suppression[3]	Growing[4]
K_1	(°C)	18°	+	13° - 15°	5°	5°
	(days)	(20)		(8 - 10)	(12)	(9)
Nakano-JA	(°C)	<18°	+	4°	5°	5°
	(days)	(20)		(12)	(6)	(10)

1 The medium (5 parts sawdust plus 1 part ricebran with 62-3% water content in 800 to 1000 ml polypropylene bottle) was inoculated with sawdust spawn after sterilization, then carried into the spawn run room.

2 Just after the spawn run ends, the inoculum and the superficial layer of media are removed. The process is named Kinkaki in Japanese.

3 Growth of fruit-bodies at the very young stage is suppressed by exposure to low temperature and a gentle air flow in order to synchronize further growth and to ensure good quality.

4 The fruit-bodies grow inside a rolled paper and are harvested at ca. 14 cm long and 140 g per bottle.

commercial cultivation system used in Japan is based on that developed by the late Mr. H. Morimoto. Since commercial cultivation was resumed in Nagano Prefecture in the 1950's, the system has undergone remarkable improvement based upon the experience of the mushroom growers. An outline of the system is shown in Table 1.

4. REPRODUCTION AND GENETICS

4.1. Mating Type, Monokaryon, Dikaryon, And Nuclear Migration

F. velutipes is a species exhibiting tetrapolar mating type, carrying A and B factors which are located on separate chromosomes. Each factor consists of linked α and β subloci such as A_α and A_β, and B_α and B_β. The map distances (recombination fraction as %) between these subloci were different with different varieties: e.g., 16.4% between B_α and B_β, 0.5 to 1.3% between A_α

and A_β (Takemaru, 1961), and 11%, 29%, 31% and 2% between B_α and B_β in the varieties T_2, T_8, K_1, and R_2, respectively. Accordingly, basidiospores carrying a new mating type factor were recovered which appeared as a result of recombination between the subloci. Recombination occurred more in B than in A. In a paired culture involving two monokaryotic mycelia of mating types which differed with respect to A and B factors, both monokaryons successfully mated to produce a dikaryon, i.e. a compatible combination. Here, hyphae from both monokaryons exchanged nuclei through hyphal bridges formed between the hyphae. The nuclei (donor nuclei) which entered the opposite monokaryotic mycelium (recipient mycelium) migrated through the mycelium, accompanied by repeated nuclear divisions. Thus, a new dikaryon was established. The rate of nuclear migration was, in some cases slower and, in other cases, faster than the rate of mycelial growth of the dikaryon. The rate of nuclear migration varied with different combinations of donor and recipient monokaryons: e.g. the nucleus of monokaryon #7-2 (Canadian origin) migrated at a rate of 0.37 cm/day through the cytoplasm of #7-1 (sister monokaryon of #7-2), and at a rate of 1.4 cm/day through #11-16 (Japanese origin) (Figure 9). Subculture of a monokaryon over a period of several months clearly caused a reduction in its capacity to drive nuclear migration irrespective of whether it served as a nuclear donor or as a nuclear recipient. Subculture over long periods finally eliminated the mating capacity of the monokaryon involved.

4.2. Interactions Between Separated Alleles Present In Conjugate Nuclei

The average growth rates of dikaryons and monokaryons in a given population were virtually identical: e.g. 0.5 cm/day on MYP medium incubated at 25°C. The growth rate was genetically determined and showed continuous variations in selfed and intercrossed progenies.

Interactions between alleles are variable. It was difficult to establish a clear relationship between the growth rates of dikaryons and those of their component monokaryons. Crossing between 'dark coloured' and 'pale coloured' fruit body types gave F_1 progenies producing fruit bodies of intermediate colour between the two parents. The major gene F and its recessive allele f of a commercial variety, Maruei, provided another example. The f allele was pleiotropic in its expression, i.e. slow mycelial growth, greater mycelial productivity, and greater fruit body productivity in homozygous condition

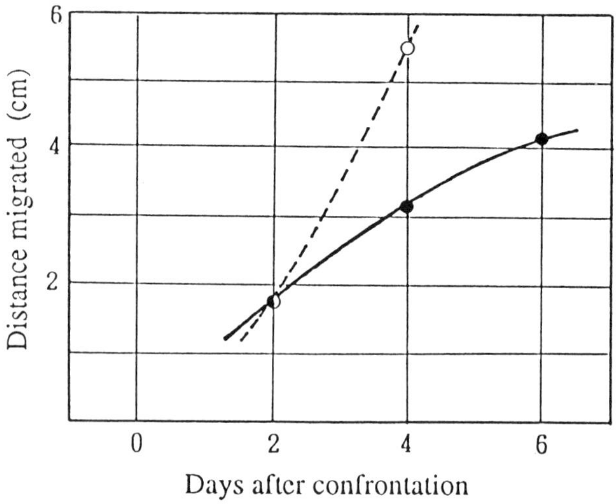

FIGURE 9. Rate of nuclear migration in *F. velutipes*.
o, nucleus #7 through cytoplasm #11
●, nucleus #7 through cytoplasm #11m.

(Kinugawa & Nakaki, 1984). The genotype f/F exhibited features which were intermediate between the f/f and the F/F genotypes with respect to the characteristics listed above (Table 2).

4.3. Monokaryotic and Dikaryotic Fruiting, and Coloration of Mycelia and Fruit Bodies

Monokaryotic fruiting ability was thought to be controlled by a set of complementary genes, since the segregation ratio (fruiting vs. non-fruiting) varied with different genotypes. Takemaru (1961) assumed that a single recessive gene was responsible for monokaryotic fruiting. In other stocks used by Kinugawa, fruiting was controlled by more than two recessive loci. Genetic analysis in the cases tested proved that there were no direct relationships between the fruiting abilities of a dikaryon and its component monokaryons (Lee and Kinugawa, 1981; 1982).

Most of the wild stocks of *F. velutipes* produce deep brown mycelia and fruit bodies, while recent commercial varieties have come to be pale coloured

TABLE 2. Average phenotypic values in association with biomass production in a selfed population of the variety Maruei.

Genotypes	Mycelial growth (mm/day)[1]	Total Production (mg/30 ml)[2]	Fruit-body yield (mg/30 ml)[2]
f/f	4.03	136.75	58.14
f/F	4.88	112.55	37.10
F/F	5.18	110.88	33.51

1, Extention rate per day on MYP.
2, Dry weight production per flask containing 30 ml. medium.

TABLE 3. Correlations of fruit-body production to total production, and to fruiting efficiency.

Population	Coefficient of Correlation[1]	
	to TP	to FE
Selfed Population of Maruei	0.553	0.577
D_1 Mixed Population (K_1mono x T_8di, T_8mono x K_1di)	0.649	0.418
D_1 Mixed Population (T_8mono x R_2di, R_2mono x T_8di)	0.598	0.795

1, Significant at 1% level
TP, Total production (fruit-body + mycelia)
FE, Fruiting efficiency
D1, Offsprings arising from di-mon mating

or white fruit body producers after continuous selection. One of the conjugate nuclei of a commercial variety, Maruei, produced coloured mycelia and monokaryotic fruiting, while the other produced colourless mycelia and non-monokaryotic fruiting. These phenotypes were segregated independently in the progenies. Colour of mycelium and fruit body was controlled by at least one single major gene with unlinked multiple genes having additive effects. Since 1981, this laboratory has undertaken a programme involving intercrossing between the varieties, Nakano-No. 1 (thick stalk) and R_2 (high yield) to select improved varieties with even high yielding ability and thicker stalks. During the selection programme, there occurred a recessive mutation which showed pleiotroic expression and which resulted in a change in the colour of the mycelium and fruit body from pale coloured to white. This mutation appeared to involve simultaneous inactivation of a series of complementary genes, since the frequencies of the segregants in mF_1 progenies (monokaryotic offsprings of a F_1 dikaryon) of the dikaryon heterozygous for mutated and unmutated genotypes deviated far from that expected of a single gene mutation. Furthermore, back mutation did not manifest the same phenotype as the original stock. 'Nakano-JA' is basically a homozygous stock of this mutation, and is the latest leading white variety in Japan (Figure 1).

5. BREEDING

5.1. Biomass and Fruiting Efficiency Determine the Yield of Fruit Body

As shown in Figure 8, total fungal production (biomass = fruit body + mycelium) was maintained at a constant level after fruiting commenced, at which time CO_2 evolution suddenly increased again. This suggested that the yield of fruit body depended upon, and was proportional to, the amount of biomass just prior to fruiting, and that the deficit in biomass due to CO_2 release was made up through additional assimilation by the mycelium. Thus, the decrease in mycelium dry weight was nearly equivalent to the dry weight increase due to fruit body formation. In intra- and intercrossed progenies cultured on the same liquid media under laboratory conditions, the variation in the yield of fruit body (F) was usually larger than that of mycelia (M), and variations in overall biomass were smallest. This last phenomenon was easily explained: i.e. maximum production by every stock culture tended to reflect a level equivalent to the amount of nutrient available in the medium under favourable laboratory conditions. We have used the term 'Fruiting Efficiency' (FE = F/(F+M)), to describe the proportion of fruit bodies produced in relation to the total biomass. The level of the total production and the FE of a particular stock was shown to be genetically determined but with some environmental impact. Accordingly, in the breeding programme, higher yielding stocks were selected from among those stocks exhibiting the highest total production and FE. During laboratory cultivation of a collection of stocks having related genotypes (e.g. selfed progenies), there appeared very high correlations between the fruit body yield and the FE and between the fruit body yield and the biomass.

5.2. Monokaryon-Monokaryon and Dikaryon-Monokaryon Mating

In basidiomycetes, crossing is possible between monokaryons compatible in mating types (mon-mon mating) and between dikaryons and monokaryons irrespective of their mating types (di-mon mating). Analysis of variance for several traits using intro- and intercross populations revealed that the commercial varieties in Japan, Maruei, K1 and R2 still have high levels of heterozygosity which can be subjected to selection. An example involving three traits of selfed progenies of Maruei (mon-mon mating) are shown in

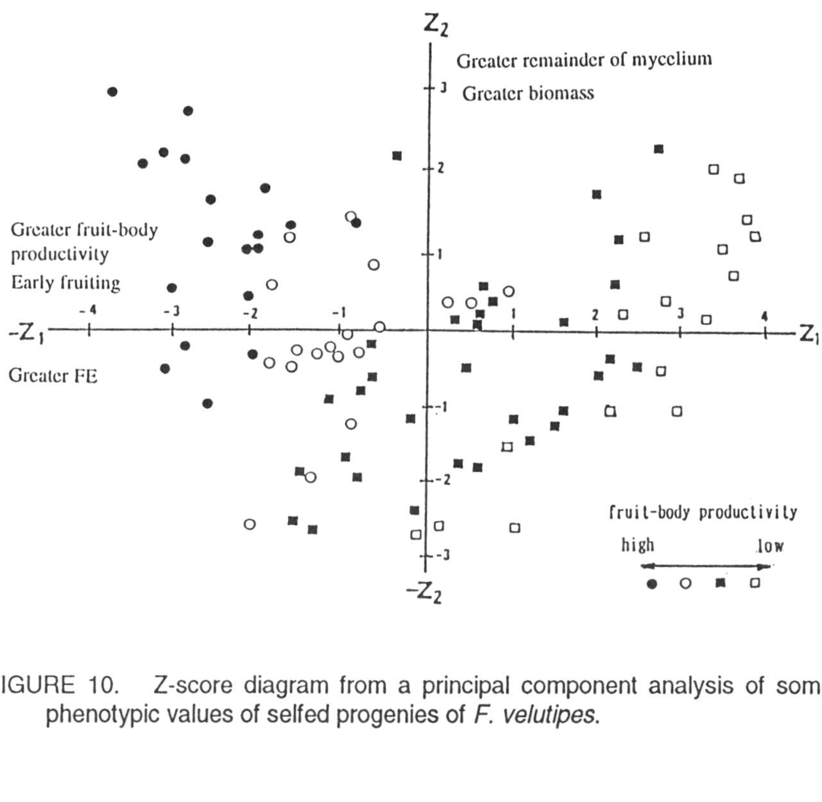

FIGURE 10. Z-score diagram from a principal component analysis of some phenotypic values of selfed progenies of *F. velutipes*.

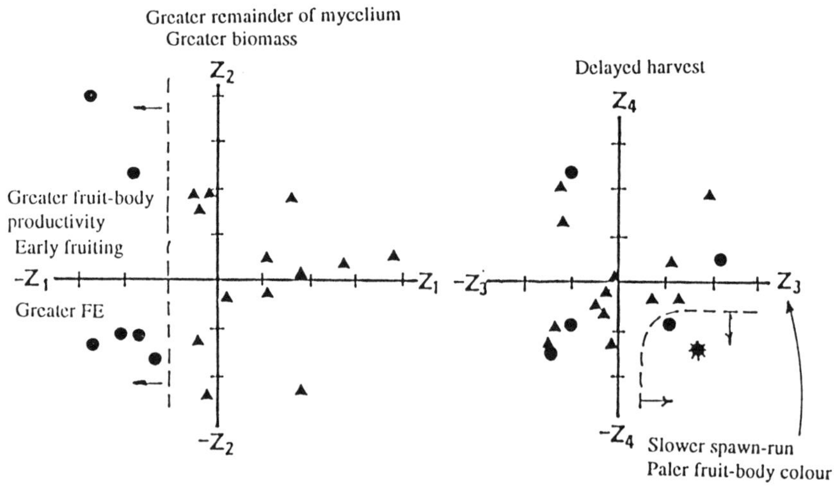

FIGURE 11. Z-score diagram and the selection of a high-yielding white variety (Nakano-JA) among selfed progenies of *F. velutipes*.
Arrows indicate the selected side. ● : good fruiter; ✸ : Nakano-JA.

Table 4, in which the genetic variance components were greater than the environmental ones, and where the variance components due to interaction were appreciable suggesting the presence of heterosis. From the result of di-mon mating shown in Table 5, the variability in donor nuclei from the parental dikaryon and of the recipient monokaryons was responsible for the variability in newly established dikaryotic offsprings (D_1). The genetic variance component was frequently of lesser significance in the dikaryotic counterpart than in the recipient monokaryons but, as in the case described in Table 5, varied with the genotypes of the parents involved. In such an inherent outbreeding organism as *F. velutipes*, searching for successful gene combinations among offspring following repeated inbreeding was very difficult because of reduced viability resulting from increased homozygosity.

5.3. Evaluation of New Dikaryotic Stocks

Efficient comparison of new dikaryotic stocks for their agronomic traits can

TABLE 4. Variance components in a selfed population of the Maruei.

Variance components	Mycelial Growth (mm/day)	Fruit-body Yield (mg/30 ml)	FE (%)
Ve (environmental)	0.35 (25%)	44.0 (30%)	22.3 (35%)
Vg (genetic)	1.04 (75%)	104.7 (70%)	40.6 (65%)
Vi (interaction)	0.18 (13%)	39.7 (27%)	17.5 (28%)
Vp (parental)	0.86 (62%)	65.0 (43%)	23.1 (37%)
Vp_1	0.40 (29%)	20.9 (14%)	5.2 (9%)
Vp_2	0.46 (33%)	44.1 (29%)	17.9 (28%)
Total	1.39	148.7	62.9

FE, Fruiting efficiency

TABLE 5. Variance components of mycelial growth in the D_1 populations arising from reciprocal di-mon matings.

Variance components	R_2(mono) x K_1(di)	K_1(mono) x R_2(di)
Ve (environmental)	0.697 (21%)	0.631 (22%)
Vg (genetic)	2.616 (79%)	2.180 (78%)
Vi (interaction)	0.551 (17%)	1.016 (37%)
Vp (parental)	2.065 (62%)	1.164 (41%)
Vp-mon	1.936 (58%)	0.428 (15%)
Vp-di	0.129 (4%)	0.736 (26%)
Total	3.313	2.811

be achieved by the use of principal component analysis (PCA). A typical example is shown in the Z-score diagram in Figure 10 using an assembly of the selfed progenies of the Maruei, in which yield ability was largely controlled by the alleles F and f (see Section 3.2). In the Z-score diagram, high yielding stocks were scattered diagonally from top middle to left below, and low yielding ones from right to left bottom. High yielding varieties exhibited the required traits, higher productivity (total production) and greater FE. For evaluation of those traits which were controlled by minor genes, the estimation of the genetic variance component was also useful (see Tables 4 and 5).

5.4. Correlations between Agronomic Traits of Dikaryotic Stocks

Relationships between some agronomic traits were identified from the correlation matrix obtained from cultivations of genetically related stocks such as F_1 and D_1 progenies. K1, Maruei, and R_2 were involved in these stocks.

In liquid media or on an agar plate medium:

(i) higher fruit body yielding stocks;
greater biomass productivity and FE,
decrease in medium pH,
wider cap,
longer length of stalk with greater coloured area,

(ii) wider cap stocks;
thicker stalk,

(iii) early fruiting stocks;
thicker stalk,

(iv) early fruiting stocks;
simultaneous harvest, etc.

In a sawdust-rice bran medium:

(i) higher yielding stocks;
simultaneous harvest,
paler cap colour,
larger cap,

uniform fruit body shape.

(ii) stocks showing simultaneous harvest;
early fruiting,
higher yielding ability,
longer stalk,
paler cap colour, etc.

Phenotypic correlations detectable under laboratory conditions were not evident under the cultivation conditions adopted in the mushroom house. Thus, stocks which grew rapidly on agar plates or in liquid media in the laboratory did not exhibit shorter spawn-running time on the SR medium in the mushroom house. There was no correlation between fruit body yields obtained in liquid media and in the SR medium. It was true, however, that parental monokaryons useful in the breeding of superior varieties were selected from those dikaryons possessing desirable traits.

5.5. Selection Procedures for Establishing New Varieties

Laboratory screening procedures were used to provide monokaryons following monobasidiospore culture, and to collect new dikaryons after crosses involving monokaryon-monokaryon or dikaryon-monokaryon matings on agar media. Vigorous dikaryons were primarily selected by comparison between their phenotypic values as measured in laboratory cultivation using SR medium contained in small bottles (180 ml). Agronomic traits were evaluated in mushroom house cultivations using ordinary culture bottles (800 to 850 ml), with an experimental design using a batch of 8 or 16 bottles for each stock with at least two replicates. Parameters for consideration and assessment included:

(a) the amount of care and handling required by each stock in terms spawn running, suppression, induction of fruiting, and growth of fruit body,
(b) quantitative evaluation of the qualities of fruit bodies, and
(c) the quantity of fruit bodies produced by each stock.

These values were analyzed by PCA and several stocks were then selected. Fig. 11 shows the Z-score diagram in the PCA, where Stock #7 was selected as the best. It was thereafter named Nakano-JA and released for commercial cultivation along with the cultivation protocol best suited to that variety. It is noteworthy that the maximum economic potential of any new variety can only be realised if the cultivation technique best suited to that variety is adopted.

REFERENCES

BANDONI, R.J. (1972). Terrestrial occurrence of some aquatic Hyphomycetes. *Canadian Journal of Botany* **50**, 2283-2288.

EGER-HUMMEL, G. (1980). Blue-light photomorphogenesis in mushrooms (Basidiomycetes). In *The Blue-Light Syndrome* p. 556. Edited by H. Senger. Berlin: Springer-Verlag.

GRUEN, H.E. (1976). Promotion of stipe elongation in *Flammulina velutipes* by a diffusate from excised lamellae supplied with nutrients. *Canadian Journal of Botany* **54**, 1306-1315.

GRUEN, H.E. (1979). Control of rapid stipe elongation by the lamellae in the fruit body of *Flammulina velutipes*. *Canadian Journal of Botany* **57**, 1131-1135.

HINTIKKA, V. & KORHONEN, K. (1970). Effect of carbon dioxide on the growth of lignicolous and soil-inhabiting Hymenomycetes. *Communicationes Instituti Forestalis Fenniae* **69**, 1-29.

KINUGAWA, K. (1977). *Collybia velutipes* can fruit under total darkness. *Transactions of the Mycological Society of Japan* **18**, 353-357.

KINUGAWA, K. & FURUKAWA, H. (1965). The fruit-body formation in *Collybia velutipes* induced by the lower temperature of one short duration. *Botanical Magazine (Tokyo)* **78**, 240-244.

KINUGAWA, K. & NAKAKI, T. (1984). A breeding method of *Flammulina velutipes*. 3: Genes regulating mycelial production. *Memoirs of the Faculty of Agriculture, Kinki University* **17**, 131-140 (Japanese with English Summary).

KINUGAWA, K & TANESAKA, E. (1990). Changes in the rate of CO_2 release from cultures of three basidiomycetes during cultivation. *Transactions of the Mycological Society of Japan* **31**, 489-500. (Japanese with English Summary).

KINUGAWA, K., TAKAMATSU, Y., SUZUKI, A., TANAKA, K. & KONDO, N. (1986). Effects of concentrated carbon dioxide on the fruiting of several cultivated basidiomycetes. *Transactions of the Mycological Society of Japan* **27**, 327-340. (Japanese with English

Summary).

LEE, P.J. & KINUGAWA, K. (1981). A breeding method for *Flammulina velutipes*. 1. Selection of monokaryotic strains by the use of testers. *Transactions of the Mycological Society of Japan* **22**, 89-102. (Japanese with English Summary).

LEE, P.J. & KINUGAWA, K. (1982). A breeding method for *Flammulina velutipes*. 2. Selection from the intercrossing and the following intracrossing. *Transactions of the Mycological Society of Japan* **23**, 177-186. (Japanese with English Summary).

LONG, T. J. (1966). Carbon dioxide effect in the mushroom *Collybia velutipes*. *Mycologia* **58**, 319-322.

NIEDERPRUEM, D.J. (1963). Role of carbon dioxide in the control of fruiting of *Schizophyllum commune*. *Journal of Bacteriology* **85**, 1300-1308.

PLUNKETT, B.E. (1956). The influence of factors of the aeration complex and light upon fruit-body form in pure cultures of an agaric and a polypore. *Annals of Botany* **20**, 563-586.

TAKEMARU, T. (1961). Genetical studies on fungi, X. The mating system in Hymenocytes and its genetical mechanism. *Biological Journal of Okayama University* **7**, 133-211.

WOOD, D.A. & FERMOR, T.R. (1985). Nutrition of *Agaricus bisporus*. In *The Biology and Technology of the Cultivated Mushroom*, p. 50. Edited by P.B. Flegg, D.M. Spencer. & D.A. Wood. Chichester: Wiley.

ZADRAZIL, F. (1978). Cultivation of *Pleurotus*. In *The Biology and Cultivation of Edible Mushrooms*, pp. 521-557. Edited by S.T. Chang. & W.A. Hayes. New York: Academic Press.

CHAPTER 6

BREEDING FOR MUSHROOM PRODUCTION IN *LENTINULA EDODES*

Albert H. Ellingboe

Department of Plant Pathology, University of Wisconsin-Madison, Madison, Wisconsin 53706, U.A.S.

1. INTRODUCTION

The Shiitake mushroom (*Lentinula (Lentinus) edodes*) is a popular item in the diet of people of Southeast Asia. It has been gaining in popularity in the U.S.A. as fresh shiitake is becoming available on a year-round basis. Traditionally, shiitake are produced on natural oak wood logs although several other species of trees are also used (Chang & Miles, 1989). However, recent years have seen an increasing interest in producing shiitake mushrooms on artificial logs composed of sawdust, bran, and a cereal grain (Chang & Miles, 1989). Advantages of the artificial log are that the time from inoculation to fructification is shorter and a higher yield of mushrooms is obtained per pound of log. The disadvantages of the artificial logs over the natural logs are the technical requirements for preparing and handling the logs to achieve the higher mushroom yields in a dependable manner.

The management strategy for producing shiitake mushrooms on artificial logs seems to be very strain dependent. Strains which were used to initiate the studies described in this paper differed greatly in several respects including the rate at which they colonized the substrate, the optimum temperature required for colonization, the time from inoculation to fruiting, resistance to contaminant fungi, the yield of mushrooms per unit dry weight of substrate, shape and color of the fruit body, and the intensity of the flavor. Strain behavior was further influenced by the composition of the substrate and the 'firmness' of the artificial log. Actual strain performance so far appears largely dependent on the management practices of the individual companies involved in mushroom production. Sharing of information about the strains used does not appear to be common practice among commercial companies in the U.S.A. However, I have the distinct feeling that the companies isolate from mushrooms sold by competitors and test these isolates under their own management practices. This is a procedure that will undoubtedly cease with the molecular procedures now available to identify individual isolates, parents and descendants (Royse & May, 1987).

Production of shiitake is not an easy endeavor. Many companies have attempted production on a commercial scale and, after failing to achieve profitable production, have subsequently gone out of business. A principal problem appears to have been difficulty in achieving the sustained production needed on a year-round basis to satisfy the fresh shiitake market. Strains may be very productive for months or years and then, without warning, exhibit greatly reduced fruiting or cease to fruit entirely. The bases for these changes are not known. Therefore, there is a need either to prevent these changes in strain productivity or to produce a large number of similar strains that can be programmed into a scheme to promote sustainable production.

2. OBJECTIVES OF THE GENETICS AND BREEDING

We began a study of the genetics and breeding of *L. edodes* in 1986 with the principal objectives of development of strains that gave high mushroom yields and were dependable in their productivity. Since most shiitake mushrooms are sliced or cut into small pieces for cooking, relatively little emphasis has been placed on the size, shape, color, etc. of the mushrooms. Attention has been given to the taste of the mushrooms, particularly whether they have a bitter aftertaste, and to the qualities relating to storage of mushrooms after harvesting. This chapter describes the way in which we proceeded to develop the breeding program. We began with an evaluation of the germplasm. Each parent dikaryon was selfed (i.e., the dikaryon was fruited, spore germination percentage determined, and single spore isolations made for subsequent matings of monosporous sibling mycelia) to determine which deleterious mutations were present, and whether strains with different accession numbers were actually identical. A standard protocol was also selected for evaluation of performance.

3. SOURCE OF PARENT CULTURES

Approximately 65 dikaryotic strains of *L. edodes* represent the basic germplasm for this program. Approximately 45 of the strains were obtained from Dr. Gary Leatham, who at that time was associated with the USDA-Forest Products Laboratory, Madison, Wisconsin. This collection included a number of strains that had been used in commercial production of shiitake mushrooms, both on synthetic logs and on natural logs. Approximately 20 strains were obtained from other individuals interested in the taxonomy, physiology, genetics, etc., of *L. edodes*. Though each of these strains has a unique accession number, it is not known how many have a common origin. Some information is available to suggest that a single isolate may have two or more different numbers. Therefore, the extent of the germplasm basis is not clear.

4. MONOKARYOTIC PROGENY

Twelve dikaryotic strains were inoculated into a medium containing, by weight, 8 parts red oak sawdust, 1 part wheat bran and 1 part rye grain. Within 1 to 3 months each of the twelve dikaryons produced one or more mushrooms. One mushroom from each strain was harvested, the stipe removed, and the cap was suspended over an agar plate. An abundant supply of basidiospores was usually available within 2 hours. The spores were suspended in sterile water and spread on 1 per cent malt agar plates. Spores germinate very slowly. Germination percentages ranged from less than 1% to about 10%. Germinated spores were transferred to 1% malt agar plates. Individual colonies that arose from single basidiospores differed greatly in growth rate, latent period before commencing growth, and other colony characteristics. Some monokaryons died after only a small colony was produced. Attempts were made to obtain at least 25 monokaryotic progenies from each dikaryon.

5. MATINGS BETWEEN MONOKARYONS

Matings were made between the monokaryons obtained from each dikaryotic strain. At least 10 *s1* monokaryons from one strain were intermated in all possible combinations. This yielded approximately 45 matings among siblings for each of the dikaryotic strains. Approximately 1/4 of the matings were expected to be sexually compatible and produce dikaryons. The expectation was based on the presumption of a bifactorial sexual incompatibility system with non-linked genes (Mori *et al.*, 1972; Tokimoto *et al.*, 1973). None of the matings among siblings from the 12 strains gave dikaryons at a frequency approaching 25%. Dikaryons were obtained in approximately 10% of the matings. This observation suggested the existence of other genes affecting mating competence in addition to the *A* and *B* mating type genes.

6. *S1* DIKARYONS

Several *S1* dikaryons obtained from each of the above crosses were also inoculated into artificial logs and carried to fruiting. Spores were collected and allowed to germinate. The percentage spore germination was much higher in this *s2* generation than in the *s1* generation. Spore germination from some *S1* dikaryons approached 100%, but was also less than 10% in spores from other *S1* dikaryons. We interpret these results as suggesting that the original dikaryotic strains contained many lethals (either mutants or gene combinations) that were lost in the *s1* generation (Royse *et al.*, 1983; Bowden *et al.*, 1991). Subsequent crosses have shown that it is almost always possible to select for dikaryons whose progeny have spore germination frequencies approaching 100%.

S1 dikaryons exhibited almost as much diversity in colony characteristics as *s1* monokaryotic progeny. The instability of certain dikaryons was associated with the instability of the dikaryotic condition. Dikaryons differed greatly in the proportion of hyphae with clamp connections. Large sectors without clamp connections occasionally occurred but were usually, although not always, reconverted to having clamp connections within a few days. Not all variability in colony characteristics was associated with dedikaryotization and redikaryotization. Sectoring for rate of linear growth occurred in some dikaryons in which essentially all hyphae had clamp connections. Therefore, the sectoring is not associated only with the stability of the dikaryotic condition.

Matings between *s2* monokaryons yielded *S2* dikaryons. The proportion of matings between *s2* monokaryons that yielded *S2* dikaryons was greater in the second generation than in the first, and approached 25%. These observations led to the conclusion that genes adversely affecting mating

	Monokaryons from dikaryon (*A1B1* + *A2B2*)			
	A1B1	*A1B2*	*A2B1*	*A2B2*
Monokaryons from dikaryon (*A3B3* + *A4B4*)				
A3B3	+	+	+	+
A3B4	+	+	+	+
A4B3	+	+	+	+
A4B4	+	+	+	+

FIGURE 1. Compatibility of matings between progenies from 2 dikaryons with different *A* and *B* alleles.

	Monokaryons from dikaryon (*A1B1* + *A2B2*)			
	A1B1	*A1B2*	*A2B1*	*A2B2*
Monokaryons from dikaryon (*A1B3* + *A3B4*)				
A1B3	-	-	+	+
A1B4	-	-	+	+
A3B3	+	+	+	+
A3B4	+	+	+	+

FIGURE 2. Compatibility of matings between progenies from 2 dikaryons that have one *A* allele in common.

	Monokaryons from one dikaryon			
	A1B1	*A1B2*	*A2B1*	*A2B2*
Monokaryons from second dikaryon				
A1B1	-	-	-	+
A1B2	-	-	+	-
A2B1	-	+	-	-
A2B2	+	-	-	-

FIGURE 3. Compatibility of matings between progenies from 2 dikaryons that share the same *A* and *B* alleles.

competency had been reduced in frequency of occurrence in the second generation.

7. CROSSING OF MONOKARYONS FROM DIFFERENT DIKARYONS

Crosses were made between *s1* monokaryotic progenies from different dikaryons. Usually 6-10 monokaryons from one dikaryon were mated in all possible combinations with 6-10 monokaryons from a second dikaryon. If the two dikaryotic parents had different mating type alleles [e.g. (*A1B1* + *A2B2*) and (*A3B3* + *A4B4*)], the proportion of matings that would give *F1* dikaryons should approach 100%. Matings would be between monokaryons *A1B1*, *A1B2*, *A2B1* or *A2B2* and *A3B3*, *A3B4*, *A4B3* or *A4B4* (See Fig. 1) (Raper 1966). Any matings that do not give dikaryons must have combinations of genes other than the *A* and *B* sexual incompatibility genes that prevent the establishment and maintenance of a dikaryon. If the two dikaryons share one *A* or *B* allele [e.g. (*A1B1* + *A2B2*) and (*A1B3* + *A3B4*)], then one fourth of the matings between the *f1* monokaryons from these two dikaryons will be sexually incompatible (Fig. 2). If two dikaryons in our collection were, in fact, of the same origin, then only one fourth of the matings between the monokaryons from the two dikaryotic stocks would be sexually compatible [e.g., (*A1B1* + *A2B2*) and (*A1B1* + *A2B2*)] (Fig. 3).

The above mating procedures have been used to determine which of the original dikaryons are distinct (and therefore may represent different germplasm), and which are either identical or have parents in common. When two of the original dikaryons were found to have all mating types in common, one was eliminated from the breeding program because it was suspected of being the same dikaryon with a different accession number.

The first set of crosses involved *s1* monokaryons from 4 original dikaryons. Approximately ten *s1* monokaryons from each original dikaryon were mated in all possible combinations with the monokaryons from the other dikaryons. Most of these matings produced dikaryons as evaluated by the presence of clamp connections. The pattern observed is presented in Fig. 1. The pattern from one mating is presented in Fig. 2. It was concluded from these data that there were 8 alleles at one sexual incompatibility locus and 7 at the other locus. Other crosses have revealed additional alleles at each locus. Present estimates are that there are more than 20 alleles at each locus (Miles & Chang, 1986). The large numbers of sexual incompatibility alleles has meant few restrictions on which monokaryons can be mated to give new *F1* dikaryons.

8. EVALUATION OF THE PRODUCTION OF MUSHROOMS BY NEW DIKARYONS

The relative yield of mushrooms from each dikaryon is expected to be dependent on the management strategy. Strains that give good yields when inoculated into natural oak logs may or may not produce similar yields when inoculated into a sawdust and grain mixture. The performance of individual dikaryons will also be affected by the temperature during colonization of the substrate, the time allowed for colonization, and the environmental conditions used to stimulate fruiting. There are obviously many variables to consider when evaluating the potential yield of good mushrooms. Clearly, it is impossible to evaluate dikaryons under an infinite series of regimes. Therefore, a set of management procedures was selected which were relatively inexpensive, and took account of the facilities and manpower resources available to the project.

Synthetic logs were chosen which contained a substrate mixture of red oak sawdust, wheat bran and rye grain in a ratio of 8:1:1. Approximately 1.5

TABLE 1. The number of dikaryotic progenies from 4 crosses producing yields within the indicated ranges.

Yield range (grams/log)	Cross A	Cross B	Cross C	Cross D
0-100	2		3	
101-200	4	2	1	
201-300	2	5	4	
301-400	12	8	0	1
401-500	14	3	5	1
501-600	6	3	4	
601-700	8	2	4	2
701-800	3		1	1
801-900			1	
901-1000			1	2

to 2.0 (average of 1.75) lbs dry weight of medium was placed in each shiitake bag (fitted with an air filter) together with 3.0-3.5 lbs water. This mixture was steamed to pasteurizing temperatures, cooled, and autoclaved 2 days later. Each log was inoculated with approximately 1 gram of spawn grown on rye grain. Normally, each dikaryon was tested initially using 3 logs. Dikaryons exhibiting high productivity in the first test were retested with 6, 12, 24 or 48 logs.

Each dikaryon was permitted to colonize the substrate within the bag for 90 days, usually at a temperature of 20-22°C. Tests were standardized at 90 days growth prior to removing the bag as a compromise between fast and slow colonizing strains. Some dikaryons had begun to fruit before the end of the 90 day period while other strains had barely colonized the substrate after this time.

At the end of the 90 day colonization period, the bags were removed and the logs placed for two weeks on shelves in rooms maintained at 18-20°C and at least 90% humidity. During this period most strains produced a bark-like surface to the log. A few strains produced a flush of mushrooms immediately upon removal of the plastic bag. Approximately two weeks after removal of the bags, the logs were soaked overnight in cold (10°C) water (Przybylowicz & Donoghue, 1988). A flush of mushrooms was usually produced within 7 days. The cold water soak was repeated every 3 weeks for a total of 3-5 flushes of mushrooms.

Dikaryons differed greatly in their responses to conditions intended to stimulate fruiting. With some dikaryons, yields of up to 60% (0.6 lbs freshweight of mushrooms for each lb of dry weight of substrate) were obtained upon opening the bag. Other dikaryons gave the largest flush of mushrooms on the fourth or fifth cold water soak. The response of a dikaryon was dependent upon its parentage.

9. YIELD OF MUSHROOMS

The frequency distribution of different yields of mushrooms among progenies from 4 crosses is given in Table 1. Crosses A & B are first generation progenies from mating monokaryons from the original dikaryons. In cross A, one parent dikaryon gave a yield of mushrooms per log of ca. 300 grams. The other parent gave a yield of ca. 400 grams per log. The mean yield of the progenies is greater than the mean of the parents. There are two obvious conclusions to draw from these data. One is that there is considerable heterosis in the crossing of monokaryotic progenies from the two parent dikaryons. The second, and probably the most likely, is that the yields produced by the parents are depressed because of the large number of lethals and other deleterious mutations that were revealed in sibling matings of progeny of most of the original dikaryons. Cross B is also of *F1* progenies and shows a distribution similar to cross A.

Crosses C and D are from mating monokaryotic progenies derived from *F1* dikaryons. There is wide variation in the yields produced by the dikaryons derived from cross C, from 0 to over 900 grams per log. Both crosses C and D gave progenies producing greater yields than crosses A and B.

10. STABILITY OF PRODUCTION

The yields produced by dikaryons from Cross A (Table 1) varied from 0 to 800 grams (0 to ca. 100%) in a rather normal distribution. The higher yields reflect, in part, a greater dependability in fruiting (i.e. greater similarity in the yield obtained from each log). Many of the strains shown as giving lower yields may produce high yields on one or two logs but very low yields on 10 other logs inoculated with the same strain. Evaluation of productivity has been based not only on maximum yields obtained from one or just a few logs but on the performance of the strain during repeated exposure to the testing regime over one or two years. The top yielding progenies from Crosses C and D (Table 1) were very consistent in their yield of mushrooms from one trial to the next. Some progenies with low or intermediate yield levels were also consistent. The yield of mushrooms in these trials is a function of both yield potential and consistency.

11. RANKING OF DIKARYONS

On average, the yields observed in the crosses presented in Table 1 are not very high. This is partly because the facilities used in this test were suboptimal. For the purpose of comparison, one or more strains used in commercial production were repeatedly included in the test cycle. In those tests reported in Table 1, the commercial strain WC305 was used as a standard and gave yields of ca. 500 grams of mushrooms per log. Though the average yield varied from one test to the next, it was reassuring to find that the relative ranking of strains that were repeatedly tested remained quite constant. When the protocols are subjected to major changes, such as promoting fruiting after only 60 days of vegetative growth, the relative ranking of the dikaryons considered most productive under the standard protocol may change markedly.

12. COMBINING ABILITY

The initial crossing was basically a diallel cross. Monokaryotic progenies from one original dikaryon were crossed with the monokaryotic progenies from each of the other original dikaryons. An attempt was made to produce a complete set of diallel crosses. The initial goal was to obtain approximately 5 monokaryons from one dikaryon that were sexually compatible with 5 monokaryons from a second dikaryon to yield a total of 25 *F1* dikaryons.

The average yield of mushrooms produced by the progenies was usually greater than the average yields of the two parents. This is probably due to the poor performance of most of the original dikaryons, most of which contained many lethals or other mutations that affected growth rate, mating competence, and/or fruiting competence.

Wide variation in yields was observed among dikaryotic progenies from a cross of monokaryotic progenies from 2 dikaryons. Most crosses gave some progenies that failed to produce any mushrooms under the conditions used for fruiting. Crosses in which the parent dikaryons shared mating types gave progeny dikaryons exhibiting smaller variations in yield, usually in the low yield range, similar to the range expected by selfing a single dikaryon.

Progeny dikaryons with top yields came from only a few crosses. Classification of the original dikaryons as good or bad parents was not clear-cut. Whether a particular parent dikaryon gave high yielding progeny was dependent on the second parent. For example, crosses involving the monokaryotic progenies of dikaryons I and II or I and III may give some high yielding *F1* dikaryons while crosses between monokaryotic progenies of II and III may not. These observations suggest that there is a major component of specific combining ability.

Among the first 300 new dikaryons tested, six gave significantly higher yields than the rest. The six came from three different crosses. It is noteworthy that five of the six had one monokaryon in common. An analysis of the performance of the siblings of this monokaryon revealed a wide range in performance when crossed with the same monokaryons from the other parents. This one monokaryon contributed to high yielding when mated with only certain other monokaryotic strains. Its siblings contributed to high yielding when mated with other monokaryons from the same dikaryons. These results also suggest a major component of specific combining ability.

The initial sets of crosses involved very few monokaryotic progeny from each parent dikaryon. The initial sets of crosses also involved dikaryons that contained many mutations affecting growth rate, lethality, colony morphology of the monokaryotic progeny, etc. In the second and third generations of crossing, there seems to be greater uniformity among the monokaryotic progeny and the dikaryotic progeny. It should be easier in subsequent generations to determine what proportion of the yield factor is attributable to general combining ability and what proportion is attributable to specific combining ability.

13. SUMMARY

The breeding program was initiated with the objectives of developing high yielding strains of *L. edodes* exhibiting dependable productivity. A standard protocol for growth and production of shiitake mushrooms on artificial logs has been adopted. Dikaryotic strains producing yields of mushrooms greater than the parent dikaryons have been developed. New strains differ greatly in their dependability of fruiting. Strains that are very consistent in their productivity have been recovered. The results to date suggest that a major component affecting increases in yield is associated with specific combining ability.

REFERENCES

BOWDEN, C.G., ROYSE, D.J. & MAY, B. (1991). Linkage relationships of allozyme encoding loci in *Lentinula edodes. Genome* **31**, 652-657.

CHANG, S.T. & MILES, P.G. (1989). *Edible Mushroom and Their Cultivation*, pp. 194-195, 201-209. Boca Raton: CRC Press.

MILES, P.G. & CHANG, S.T. (1986). The collection and conservation of gene of *Lentinula.* In *Cultivating Edible Fungi*, pp. 227-234. Edited by P.J. Wuest, D.J. Royse, & R.B. Beelman. Amsterdam: Elsevier Publishers.

MORI, K., ZENNYOJI, A. & KUGIMIYA, N. (1972). Analysis of the incompatibility factors in natural population of *Lentinula edodes. Japanese Journal of Genetics* **47**, 359.

PRZYBYLOWICZ, P. & DONOGHUE, J. (1988) *Shiitake Growers Handbook*, p. 151. Dubugue: Kendall-Hunt Publishing Company.

RAPER, J.R. (1966). *Genetics of Sexuality in Higher Fungi*, p. 51. New York: Ronald Press.

ROYSE, D.J., SPEAR, M.C. & MAY, B. (1983). Single and joint segregation of biochemical loci in the shiitake mushroom, *Lentinula edodes. Journal of General and Applied Microbiology* **29**, 217-222.

ROYSE, D.J. & MAY, B. (1987). Identification of shiitake genotypes by multilocus electrophoresis: catalog of lines. *Biochemical Genetics* **25**, 705-716.

TOKIMOTO, K., KOMATSU, M. & TAKEMARU, T. (1973). Incompatibility factors in the natural population of *Lentinula edodes* in

Japan. *Report of the Tottori Mycological Institute (Japan)* **10**, 371-376.

CHAPTER 7

PROTOPLAST TECHNOLOGY AND EDIBLE MUSHROOMS

John F Peberdy and Hilary M Fox

Department of Life Science, University of Nottingham,
Nottingham NG7 2RD, United Kingdom.

1. INTRODUCTION

Commercial and marketing pressures have been important driving forces in the development and expansion of the established areas of fungal biotechnology. Central to the improvement of a fermentation process, be it the production of an antibiotic in a liquid system or a solid state system such as used in mushroom production, has been the improvement of the organism. Traditionally in the fermentation industry this has been based on mutation and selection of improved strains; breeding techniques involving recombination have been less important. In mushroom production the very opposite is the case, as the product itself represents the climax of sexual development and recombination through meiosis provides the most opportunity to develop new strains.

About 20 years ago strain improvement practice for commercially important fungi was given a new impetus with the realisation of a new area of technology based on isolated protoplasts. Enzymes, of microbial origin, were discovered which digested away the fungal cell wall liberating the hyphal protoplasts as discrete units of the whole provided some external osmotic support was available. Of major importance was the observation that when protoplasts are supplied with suitable nutrients they resynthesize a new cell wall and undergo a process of regeneration to re-establish the hyphal form. The cell wall has many roles including that of acting as a barrier protecting the protoplast (Peberdy, 1989). Without this barrier it was possible to explore a whole range of new manipulations of these naked cells many of which had important genetic consequences. The first step was the development of procedures for protoplast fusion to be followed about 5 years later with methods for transformation. This review will be concerned with the technology associated with protoplast fusion which is highly relevant to the improvement of commercially important mushrooms. More general reviews on the subject published most recently are Peberdy (1990; 1991).

2. THE PROTOPLAST SYSTEM IN BASIDIOMYCETES

The techniques of protoplast isolation have been described for a large number of basidiomycete species, many of which are cultivated on a commercial scale (Table 1).

2.1. Mycolytic Enzymes

A wide range of wall degrading enzymes have been used in the release of protoplasts from many edible species. Information on wall composition in this group of fungi is very restricted (Wessels & Seitsma, 1979; Mendoza *et al.*, 1987; Hernandez *et al.*, 1990) and whilst chitin and β-glucan are the major polymers, it is likely that the minor components are also of some significance to the process of wall degradation.

A range of enzymes are available from several different suppliers (Hocart & Peberdy, 1989) which include a number of cellulase and other β-1,3-D-glucanases which would have the main destructive effect on the cell wall (Hamlyn *et al.*, 1981; Hocart *et al.*, 1987; Yu & Chang, 1987). The experience of most workers is that mixtures of these complex preparations are frequently better than one product alone; the reasons for this significant effect are unclear.

2.2. The Organism

TABLE 1. Protoplast Isolation Systems Employed for Various Basidiomycete Species.

ORGANISM	STABLILIZER/ pH	LYTIC ENZYMES USED*	REFERENCE
Agaricus bisporus	0.5M $MgSO_4$/ 5.8	11	De Vries & Wessels (1973)
(=*A. brunnescens*)	0.5M $MgSO_4$/ 5.9	15	Anderson *et al.* (1984)
	0.6M $MgSO_4$/ 5.8	13	Chang *et al.* (1985)
	Not Given	1,15,16	Yoo *et al.* (1985)
	0.6M $MgSO_4$/ 5.0	15	Castle *et al.* (1987)
	0.6M Sucrose/ Not Given	11	Sonnenberg *et al.* (1988)
	0.5M $MgSO_4$, 1mM $CaCl_2$	15,8	Mendoza *et al.* (1991)
A. bitorquis	0.6M Sucrose/ Not Given	11	Sonnenberg *et al.* (1988)
A. campestris	0.5M $MgSO_4$/ 5.8	11	De Vries & Wessels (1973)
Agrocybe aegerita	Not Given	15	Noel & Labarere (1989)
Auricularia auricula	0.6M $MgSO_4$/ 5.8	13	Chang *et al.* (1985)
A. polystricta	0.5M Mannitol/ 5.5	2,5	Ohmasa *et al.* (1987)
Cenococcum geophilum	0.5M Mannitol/ 5.6	8,15	Barrett *et al.* (1989)
Coprinus bilantus	0.5M Mannitol/ 5.5	6,8	Burrows *et al.* (1990)

TABLE 1 continued.

ORGANISM	STABLILIZER/ pH	LYTIC ENZYMES USED*	REFERENCE
C. cinereus	0.5M $MgSO_4$/ 5.8	11	De Vries & Wessels (1973)
	1.0M Sucrose/ Not Given	8	Moore *et al.* (1975)
	0.5M $MgSO_4$/ 5.8	6	Akamatsu (1983)
	0.5M $MgSO_4$/ 5.8	5,6,8	Morinaga *et al.* (1985)
	0.55M $MgSO_4$/ 5.6	11	Kitamoto *et al.* (1987)
	1.2m $MgSO_4$/ 5.5	11	Kitamoto *et al.* (1988)
C. macrohizus	0.5M Mannitol/ 5.5	6,8,18	Kiguchi & Yanagi (1985)
	0.5M Mannitol/ 5.5	6,8,17	Yanagi & Takebe (1984)
	0.5M Mannitol/ 5.5	6,8,18	Yanagi *et al.* (1985)
C. pellucidus	0.5M $MgSO_4$/ 5.8	2,6,8,17	Morinaga *et al.* (1985)
C. phlystidosporus	1.2M $MgSO_4$/ 5.5	11	Kitamoto *et al.* (1988)
Flammulina velutipes	0.6M Sucrose/ Not Given	2,6,8,17	Yamada *et al.* (1983)
	1.2M $MgSO_4$/ 5.5	12	Kitamoto *et al.* (1984)
	0.5M Mannitol/ 5.5	5,9,14,17	Ohmasa *et al.* (1987)
	1.2M $MgSO_4$/ 5.5	11	Kitamoto *et al.* (1988)
	0.6M Sucrose/ 6.2	3,15	Yea *et al.* (1988)

TABLE 1 continued.

ORGANISM	STABLILIZER/ pH	LYTIC ENZYMES USED*	REFERENCE
Flavolus arcularius	1.2M $MgSO_4$/ 5.5	11	Kitamoto *et al.* (1988)
Gandoderma applanatum	Not Given	2,6,15	Yoo (1989)
G. lucidum	0.5M $MgSO_4$/ 5.8	11	De Vries & Wessels (1973)
	0.5M Mannitol/ 5.5	6,8,18	Yanagi *et al.* (1985)
	0.6M Sucrose/ 5.8	2,15	Shin *et al.* (1986)
	0.6M Sucrose/ 5.8	2,15	Choi *et al.*. (1988)
	0.6M Sucrose/ Not Given	2,15	Um *et al.* (1988)
Grifola frondosa	0.5M Mannitol/ 5.5	6,8,18	Yanagi *et al.* (1985)
	0.5M Mannitol/ 5.5	5,9,14,17	Ohmasa *et al.* (1987)
Hebeloma circinans	0.5M Mannitol/ 5.6	15	Barrett *et al.* (1989)
H. cylindrosporum	0.5M Mannitol/ 5.6	15	Barrett *et al.* (1989)
Hypsizygus marmoreus	0.5M Mannitol/ 5.5	5,9,14,17	Ohmasa *et al.* (1987)

TABLE 1 CONTINUED.

ORGANISM	STABLILIZER/ pH	LYTIC ENZYMES USED*	REFERENCE
Laccaria bicolor	0.5M Mannitol/ Not Given	15	Kropp & Fortin (1985)
	0.5M Mannitol/ 5.6	15	Barrett *et al.* (1989)
L. laccata	0.5M Mannitol/ 5. 6	15	Barrett *et al.* (1989)
Lentinula edodes	0.5M $MgSO_4$/ 5.8	11	De Vries & Wessels (1973)
(=*Lentinus edodes*)	0.5M $MgSO_4$/ 5.8	4,10,12	Ushiyama & Nakai (1977)
	1.2M $MgSO_4$/ 5.5	11	Kitamoto *et al.* (1984)
	0.5M $MgSO_4$/ 5.8	6,10,17	Toyoda *et al.* (1984)
	0.6M $MgSO_4$/ 5.0	2,5	Woo & Yoon (1985)
	0.5M Mannitol/ 5.5	6,8,18	Yanagi *et al.* (1985)
	0.6M Mannitol/ 4.6	5,8	Kawasumi *et al.* (1987)
	0.5M Mannitol/ 5.5	2,5,9	Ohmasa *et al.* (1987)
	1.2M $MgSO_4$/ 5.5	11	Kitamoto *et al.* (1988)
	0.6M $MgSO_4$/ 5.6	2,6,8,20	Koga *et al.* (1988)
Lentinus tigrinus	0.5M $MgSO_4$/ 5.8	11	De Vries & Wessels (1973)

TABLE 1 continued.

ORGANISM	STABLILIZER/ pH	LYTIC ENZYMES USED*	REFERENCE
Lyophyllum shimeji	0.6M $MgSO_4$/ 5.6	2,6,17	Abe *et al.* (1984)
	0.6M $MgSO_4$/ 5.6	2,6,8,20	Koga *et al.* (1988)
L. ulmarium	0.6M Mannitol/ 5.5	6,8,18	Yanagi *et al.* (1984)
Oudemansiella mucida	0.7M Glucose/ Not Given	16	Homolka *et al.* (1988)
Phanerochaete chrysosporium	0.6M $MgSO_4$/ 5.5	3,15	Gold *et al.* (1983)
	0.7M $MgSO_4$/ 5.9	15	Gaskell *et al.* (1991)
Pholiota nameko	0.5M Mannitol/ 5.5	6,8,18	Yanagi & Takebe (1984)
	0.5M Mannitol/ 5.5	2,5	Ohmasa *et al.* (1987)
	1.2M $MgSO_4$/ 5.5	11	Kitamoto *et al.* (1988)
Pisolithus tinctorius	0.5M Mannitol/ 5.6	15	Barrett *et al.* (1989)
Pleurotus columbinus	0.7M Mannitol/ Not Given	2,5,9	Toyomasu & Mori (1987)

TABLE 1 continued.

ORGANISM	STABLILIZER/ pH	LYTIC ENZYMES USED*	REFERENCE
P. cornucopiae	0.6M Mannitol/ 5.6	2,5,8,17	Wakabayashi *et al.* (1985)
	0.5M Mannitol/ 5.5	5,9,14,17	Ohmasa *et al.* (1987)
P. corticatus	0.5M $MgSO_4$/ 5.8	11	De Vries & Wessels (1973)
P. cystidiusus	0.5M Mannitol/ 5.5	5,9,15,17	Ohmasa *et al.* (1987)
P. florida	0.6M $MgSO_4$/ 5.8	1,15,16	Yoo *et al.* (1984)
P. ostreatus	0.6M Mannitol/ Not Given	2,5,8,17	Yamada *et al.* (1983)
	0.4M $MgSO_4$/ 5.5	13	Chang *et al.* (1985)
	Not Given	15	Go *et al.* (1985)
	0.5M Mannitol/ Not Given	2,5,8,19	Yanagi *et al.* (1985)
	0.7M Mannitol/ Not Given	2,5,9	Toyomasu *et al.* (1986)
	0.5M Mannitol/ 5.5	2,5,8,19	Ijima & Yanagi (1986)
	1.2M $MgSO_4$/ 5.5	11	Kitamoto *et al.* (1988)
	0.5M Mannitol/ 5.5	8,15	Ohmasa *et al.* (1987)
	Not Given	1,2,15	Yoo *et al.* (1989)

TABLE 1 CONTINUED.

ORGANISM	STABLILIZER/ pH	LYTIC ENZYMES USED*	REFERENCE
P. pulmonaris	0.7M Mannitol/ Not Given	2,5,9	Toyomasu & Mori (1987)
P. sajor-caju	0.4M $MgSO_4$/ 5.5	13	Chang *et al.* (1985)
	Not Given	15	Go *et al.* (1985)
	0.6M $MgSO_4$/ 5.0	3,15	Lau *et al.* (1985)
	0.7M Mannitol/ Not Given	2,5,9	Toyomasu & Mori (1987)
P. salmoneo-stramineus	0.7M Mannitol/ Not Given	2,5,9	Toyomasu *et al.* (1986)
	0.5M Mannitol/ 5.5	5,9,14,17	Ohmasa *et al.* (1987)
	0.6M Sucrose/ Not Given	1,2,15	Yoo *et al.* (1989)
P. sapidus	0.6M Sucrose/ 6.0	1,2,15	You *et al.* (1988)
Polystictus versicolor	12.5% Sucrose/ 6.5	16	Strunk (1965)
Rhizoctonia solani	0.6M Mannitol/ 5.2	2,6,14	Hashiba & Yamada (1982)
Robillarda sp.	0.6M Mannitol/ 6.0	2,5,8,20	Kuwabarda *et al.* (1989)

TABLE 1 CONTINUED.

ORGANISM	STABLILIZER/ pH	LYTIC ENZYMES USED*	REFERENCE
Schizophyllum commune	0.6M $MgSO_4$/ 5.8	11	De Vries & Wessels (1972)
	0.5M $MgSO_4$/ 6.75	7,15	Munoz-Rivas *et al.* (1986)
	1.2M $MgSO_4$/ 5.5	11	Kitamoto *et al.* (1988)
Sclerotium rolfsii	0.6M KCL/ 5.0	15	Kelkar *et al.* (1990)
Sistotrema brinkmannii			Anderson & Cendese (1984)
Termitomyces clypeatus	0.5M KCL/ 6.0	3,8,15	Mukherjee & Sengupta (1988)
Trametes sanguinea	1.2M $MgSO_4$/ 5.5	11	Kitamoto *et al.* (1988)
Tremella fuciformis	0.5M Mannitol/ 5.5	6,8,18	Yanagi *et al.* (1985)
Tricholoma matsutake	0.6M $MgSO_4$/ 5.6	2,6,17	Abe *et al.* (1982)
	0.5M Mannitol/ 5.5	5,9,14,17	Ohmasa *et al.* (1987)
	0.6M $MgSO_4$/ 5.6	6,2,8,20	Koga *et al.* (1988)

TABLE 1 continued.

ORGANISM	STABLILIZER/ pH	LYTIC ENZYMES USED*	REFERENCE
Volvariella bombycina	Not Given	2	Stille (1984)
	Several	15	Chang *et al.* (1985)
Volvariella volvacea	0.5M $MgSO_4$/ 5.8	11	De Vries & Wessels (1973)
	0.6M $MgSO_4$/ 5.8	15	Hamlyn *et al.* (1981)
	1.2M $MgSO_4$/ 5.8	11,15	Santiago (1982a)
	1.2M KCL/ 5.8	11,15	Santiago (1982b)
	Not Given	2,15,16	Yoo *et al.* (1985)
	0.6M NaCL/ 6.0	15	Mukherjee & Sengupta (1986)

* KEY TO ENZYME CODES: 1 = β-D-Glucanase; 2 = β-Glucuronidase; 3 = Cellulase CP; 4 = Cellulase CP-1500; 5 = Cellulase "Onazuka" RS; 6 = Cellulase "Onazuka" R10; 7 = Cellulase t.v.; 8 = Chitinase; 9 = Driselase; 10 = Helicase; 11 = Induced Lytic Enzyme; 12 = Lytic Enzyme 3; 13 = Lywallzyme; 14 = Macerozyme R10; 15 = Novozym 234; 16 = Snail Enzyme; 17 = Zymolase 5,000; 18 = Zymolase 60,000; 19 = Zymolase 100,000; 20 = Zymolase 20T.

The physiological age and hence the condition of mycelium obtained is another important factor in protoplast formation. In fast growing species mycelium is grown up to the early linear phase. However, in many basidiomycetes growth is very slow and the different phases of growth are less distinct. Most reports indicate poorer protoplast yields in these fungi compared to non-basidiomycete species (Peberdy, 1989) and mycelial age could be a factor. A further technical problem with these fungi is the inoculum and achieving one that is standardized. In some instances, e.g. *Phanerochaete chrysosporium*, germinating basidiospores to produce young mycelium was effective for protoplast production (Gold *et al.*, 1983). Short pre-germination of basidiospores from produced useful material for good protoplast production in *Schizophyllum commune* (Ullrich *et al.*, 1985).

Because of the absence of asexual spores in most basidiomycetes and the variable germination of sexual spores, macerated hyphae are most commonly used as inoculum for cultures to produce mycelium. This results in a high degree of variability because the hyphal fragments are heterogeneous with respect to age and viability. (Abe *et al.*, 1982; Anderson *et al.*, 1984; Morinaga *et al.*, 1985; de Vries & Wessels, 1972). In principle, spores are an ideal source of protoplasts but are generally more resistant than vegetative cells to the lytic enzymes. In the basidiomycetes the oidia from *Coprinus cinereus* (Akamatsu *et al.*, 1983) have proved to be the only example of the use of spores. In the original report an incubation period of 24 hours was necessary to maximize protoplast production. However, other workers have recently made improvements to the procedure (Mellon *et al.*, 1987).

Following the practice first established in *Saccharomyces cerevisiae*, several workers have demonstrated the value of pre-treatment of mycelium prior to lytic digestion. The materials used for this treatment have been reducing agents including dithiothrietol and mercaptoethanol (Davis, 1985). The relevance of such treatment for mycelium of basidiomycetes has probably not been rigorously investigated, there being only one report of its application (Choi *et al.*, 1987).

2.3. Osmotic Stabilizers

Survival of protoplasts liberated from hyphae following enzymic digestion depends on osmotic support provided by the digestion medium. In general, inorganic salts, sugars and sugar alcohols have proved to be best for filamentous fungi (Peberdy, 1979; Davis, 1985). Amongst the basidiomycetes surveyed, $MgSO_4$, mannitol and sucrose have been the most widely used (Table 1).

Determination of the most effective osmoticum for a given fungus is quite empirical and the best compound for one species may not be the best for another. This underlies the ongoing situation in protoplast technology, namely the events and processes involved are not understood despite the fact that the first reports on the technique were published some 30 years ago (Emerson & Emerson, 1958).

An important interaction in the lytic digestion mixture concerns the osmotic stabilizer and the enzymes of the complex. Yu and Chang (1987) observed that chitinase was most inhibited by the various compounds tested although this is probably variable depending on the specific enzyme and its source.

Other compounds have been adopted as physiological stabilizers and included in the lytic digestion solution. These are normally inorganic salts, e.g. $CaCl_2$ and $MgSO_4$ and are used at lower concentrations in the range 1-100mM (Thomas & Davis, 1980). These compounds are believed to aid the stability of the protoplast membrane. However, there are no reports to date of their use with basidiomycetes.

2.4. Cell Wall Formation and Protoplast Regeneration

The process of new cell wall formation on the protoplast surface and regeneration to the normal cell form are key events in the application of protoplasts in genetic manipulation. Details of the process of the new wall formation are not relevant to this overview and can be found elsewhere (de Vries & Wessels, 1975; van der Walk & Wessels, 1976). However, what is important is the frequency of regeneration which is normally assessed on the basis of colony forming units. Generally, such frequencies for basidiomycetes are low (Peberdy, 1989) and may present problems for genetic manipulation.

As with protoplast isolation, the optimization of conditions for regeneration is also empirical with factors such as the use of agar overlays, agar concentration, osmotic stabilizer and sometimes specific medium ingredients being taken into account. Whilst some workers have favoured the inoculation of basidiomycete protoplasts onto an agar surface (Hashiba & Yamada, 1982; Kiguchi & Yanagi, 1985; Lau *et al.*, 1985) and others agar overlays (Abe *et al.*, 1984; Santiago, 1981; Yamada *et al.*, 1983), where comparisons have been made no significant benefits of either method has been shown (Magae *et al.*, 1985; Morinaga *et al.*, 1985). On the contrary Gold *et al.* (1983) showed that agar concentration in the overlay medium was important in maximizing regeneration in *Phanerochaete chrysosporium.*

Where comparisons can be made the published data on the effect of osmotic stabilizers on regeneration is also equivocal.

In their work with *Pleurotus sajor-caju*, Lau *et al.* (1985) observed that mannitol was the most effective osmoticum in the regeneration medium, but Abe *et al.* (1984) had earlier shown this compound to be inferior to inositol.

The enzyme(s) used for protoplast isolation may also affect the regenerative potential of the protoplasts. For example, several workers have described a detrimental effect on regeneration of protoplasts from several basidiomycetes which correlated with the use of the Novozym 234 complex (Yanagi *et al.*, 1985; Sonnenberg *et al.*, 1988). Other reports (Wakabayashi *et al.*, 1985) suggest that a prolonged incubation in the enzyme solution increased the viability of *Pleurotus cornucopia* protoplasts.

Several workers have attempted to improve the regeneration frequencies of protoplasts from a range of basidiomycete species by the addition of various supplements to the basal medium. In part these additives reflect the natural substrate for the growth of these fungi and include materials such as compost extract and sulphide pulp waste (Ijima & Yanagi, 1986; Kawasumi *et al.*, 1987).

3. MUTAGENESIS OF PROTOPLASTS

Traditional fungal genetics has been based on the generation of mutants as isogenic strains derived from a specific progenitor. In a basidiomycete this would be done from a monokaryon, which in most cases do not produce suitable spores. In this situation hyphal fragments would be used. However, protoplasts can prove to be a reasonable alternative. Thus, mutagenesis of protoplasts from a *Volvariella volvacea* monokaryon gave rise to both auxotrophic and morphological mutants (Mukherjee & Sengupta, 1986). Auxotrophic mutants have also been obtained, via protoplasts, from *Pleurotus* spp. (Toyamasu *et al.*, 1986; Toyamasu & Mori, 1987a) and from *Oudemansiella mucida* (Homolka *et al.*, 1988).

4. PROTOPLASTS AND DE-DIKARYOTIZATION

With the exception of *Agaricus bisporus* the generation of new strains can be achieved by making crosses between different compatible monokaryons. The isolation and regeneration of protoplasts from a dikaryon provides another approach to the disassociation of the component nuclei and has been applied in a range of species.

5. PROTOPLASTS AND GENETIC MANIPULATION

Protoplasts are used in two areas of genetic manipulation, fusion and transformation. Through protoplast fusion it is possible to bring together whole genomes of related (isogenic) or non-related strains and even different species and promote recombination leading to the production of novel phenotypes. Protoplast fusion, therefore, can provide a mechanism to overcome natural incompatibility barriers that exist between different strains and species. Once the fusion event has been achieved and the protoplasts regenerate, then the events which follow are those associated with normal sexual or parasexual reproduction.

Transformation is the approach used to manipulate organisms at the level of individual genes, the introduction of the gene of interest into the cell being mediated by a vector DNA molecule. To date the most successful method for the introduction of DNA molecules into fungal cells has depended on the use of protoplasts. However, other approaches are being developed and slowly exploited in fungi along with other organisms.

5.1. Protoplast Fusion - The Technique

5.1.1. Fusogenic methods. The most widely adopted method to produce fusion events between protoplasts has been the use of polyethylene glycol (PEG) in the presence of $CaCl_2$. This situation is true for basidiomycetes as it is for all other fungi. Also, there have been two reports on the use of electrofusion with *Agaricus bisporus* (Sonnenberg & Wessels, 1987) and with *Pleurotus ostreatus* (Magae *et al.*, 1985). Fusion events brought about by PEG are very random, in contrast to electrofusion where more control can be achieved. However, this technique is not without its limitations not least the determination of optimal conditions to achieve fusion.

The PEG-Ca^{++} system is not fully understood in terms of the mechanism of the fusion event. Calcium ions clearly play a critical role, as has been well demonstrated in experiments with *Coprinus macrorhizus* (Kiguchi & Yanagi, 1985). Abe *et al.* (1982) also described the enhancing effect of Ca^{2+} ions on protoplast fusion in *Tricholoma matsutake*.

Observation of aggregates of protoplasts after PEG treatment normally reveal the formation of heterogeneous masses. This was the case for *Coprinus macrorhizus* (Kiguchi & Yanagi, 1985). PEG can have damaging effects on protoplasts so reducing viability. Abe *et al.*, (1982) described this situation in *Tricholoma matsutake.*

5.1.2. Selection strategies. Although recombinants have been recovered from a protoplast fusion cross in *Penicillium chrysogenum* without the introduction of selective markers into the two strains involved (Lein, 1986), for most published work a strategy for selection has been adopted. In the case mentioned the resources were available to screen many thousands of fusant progeny, a necessity in this situation. Selection based on some pre-existing natural difference or introduced genetic difference in the two strains is more precise and time-saving.

The selection strategies first adopted in protoplast fusion crosses were based on approaches that had been extensively used in the classical *Aspergillus* genetics to promote the establishment of balanced heterokaryons. These require the introduction of "tight" complementary auxotrophic mutations in the strains enabling the recovery of products, following PEG-induced fusion or electrofusion, on a minimal medium. Because auxotrophy is known to affect other aspects of metabolism, such marker genes are not adopted in organisms used in industrial fermentations e.g. antibiotic and enzyme production. Furthermore, the introduction of mutations in production strains in order to facilitate a cross also raises the possibility of introduction of mutations detrimental to product yield. Whether a similar situation exists in basidiomycetes is unknown.

To overcome this problem alternative approaches to the introduction of useful selective markers have been developed. One such type of marker is a carbon or nitrogen substrate non-utilisation mutation which can be introduced as a consequence of resistance to a particular growth inhibitor. Selection can also be based on resistance to growth inhibitors including the compounds which are marketed as fungicides (Table 2).

5.2. Applications of Protoplast Fusion in Basidiomycetes

In several fungi protoplast fusion has proved to be central to the basic procedures for classical genetics. In the basidiomycetes the most likely application is in the generation of novel interstrain or interspecies hybrids. The importance of the complex mating type and incompatibility mechanisms that operate in these fungi in relation to the generation of such hybrids have still to be resolved.

5.2.1. Interstrain crosses. In the several published reports on protoplast fusion in basidiomycetes, e.g. *Phanerochaete chrysosporium* (Gold *et al.*, 1983), *Coprinus macrorhizus* (Kiguchi & Yanagi, 1985), *Pleurotus ostreatus* (Ohmasa, 1986), *Pleurotus salmoneo-stramineus* (Toyomasu & Mori, 1987a) and *Lentinus edodes* (Kawasumi *et al.*, 1987), selection of fusion products was achieved by complementation of auxotrophs in the strains crossed. In most reports, heterokaryons (dikaryons) were recovered their nature being ascertained by protoplast formation and characterization of the parental genotypes in monokaryotic regenerants (Gold *et al.*, 1983) and fruiting (Toymasu & Mori, 1987).

Protoplast fusion crosses involving mating type compatible strains of *Coprinus macrorhizus* (Kiguchi & Yanagi, 1985) gave rise to heterokaryons which fruited normally. Heterokaryons were also obtained at a similar frequency when protoplasts from incompatible strains were fused, but not surprisingly these failed to develop fruit bodies.

Ohmasa (1986) described the formation of an interstrain hybrid of *Pleurotus ostreatus* by protoplast fusion which had characteristics related to fruit body formation, morphology and environmental triggers for primordia development that are intermediate between the parental strains. A range of progeny, showing variation with respect to growth rate, was also obtained from crosses between complementary auxotrophs of *Pleurotus salmoneo-stramineus* (Toymasu & Mori, 1987a). Such variability might be expected in the progeny of protoplast fusion crosses compared to hyphal anastomosis crosses. The monokaryotic basidiomycete mycelium can be expected to be heterokaryotic with respect to nuclear genes, other than mating type. Random fusions of uninucleate protoplasts of differing genotypes would therefore

TABLE 2. Minimium inhibitory concentrations of selected antifungal agents on several *Pleurotus* species.

SPECIES	MIC (μg. ml-1)								
	Hy	Ph	5-FI	Cx	Cy	Pz	Fz	Iz	
P. colombinus	-	40	-	0.6	10	60	200	20	
P. florida		65	80	-	0.4	8	150	140	60
P. ostreatus	-	40	5	1	20	20	80	150	
P. pulmonarius	60	>80	-	1.2	20	60	20	30	
P. sapidus		-	-	-	1.2	8	50	200	200
P. sajor-caju	55	55	8	0.8	30	160	180	30	

Key to agents: Cx = Carboxin; Cy = Cycloheximide; 5-FI = 5-Fluroindole; Fz = Flurilazole; Hy = Hygromycin B; Ph = Pheomycin; Iz = Imazalil; Pz = Propiconazole.
(Jia Jian-hua & Peberdy, J.F., unpublished).

generate wider genome variability than would arise through hyphal anastomoses.

5.2.2. Interspecific crosses. To date, attempts at hybridization between basidiomycetes have been limited to a few species of *Pleurotus* and *Ganoderma*. Whilst fusants were obtained in several crosses, fruit body formation has been observed in only two hybrids, *Pleurotus ostreatus* x *florida* (Yoo *et al.*, 1984) and *Pleurotus ostreatus* x *columbinus* (Toyomasu & Mori, 1987a). In the former case, the fruit bodies produced few if any spores. Clamp connections were found in fusants that developed fruit bodies. Incompatibility between *Pleurotus ostreatus* and *Pleurotus salmoneo-stramineus* was overcome by induced protoplast fusion between auxotrophic monokaryons (Toyomasu *et al.*, 1986). Further crosses between *Pleurotus ostreatus*, *Pleurotus pulmonaris* and *Pleurotus sajor-caju* have been attempted (Toyomasu & Mori, 1987a). However, of the six possible combinations only four yielded fusion products. The crosses that failed were *Pleurotus columbinus* x *Pleurotus pulmonaris* and *Pleurotus ostreatus* x *Pleurotus pulmonaris*. Hybrids obtained from all the other crosses differed in terms of colony morphology and growth rate (Toyamasu & Mori, 1987b). The hybrid of *Pleurotus ostreatus* x *Pleurotus columbinus* produced fruit bodies on sawdust medium but the rest failed to do so.

Protoplast fusion products were also reported from crosses between *Ganoderma lucidum* and *G. applanatum* (Park *et al.*, 1988; Um *et al.*, 1988). On a rich growth medium the fusants segregated to form sectors of hyphae with and without clamp connections. The loss of selective pressure provided by a minimal medium was apparently necessary to maintain the heterokaryotic state.

Clearly, it is of interest to understand the underlying interaction of mating type genes in these hybrids. If the genes and their products may be highly conserved then the potential for interaction is conceivable and the barrier to "natural" hybridization may therefore lie at the cell wall through the lack of recognition. More fundamental, however, is the taxonomic distinction of the species used in these crosses. This is particularly the case with *Pleurotus* species where it appears that designation of several supposed species is rather doubtful (Buchanan P., personal communication) suggesting that some of the crosses described above may in fact be interstrain and not interspecies.

5.3. Molecular Markers and the Assessment of Interspecies Progeny

In these as in all interspecies crosses, the availability of further supporting evidence both from the repetition of the crosses and a broad genetical/ biochemical analysis of the fusion products is essential. Isoenzyme profiles have been shown to be effective molecular markers in basidiomycetes (Royse & May, 1982a,b; Royse *et al.*, 1983a,b; Royse *et al.*, 1987) and have been used in the analysis of crosses in several species (Toyomasu & Mori, 1987b).

Restriction fragment length polymorphisms (RFLPs) provide a more reliable molecular marker. To date, this technology has been used to address the question of cell line authentication and patenting specific cultivars (Horgen & Anderson, 1987) The potential of the technique to discriminate between strains of different origins has been clearly demonstrated in *Coprinus cinereus* (Wu *et al.*, 1983). This latter work indicates the potential of RFLPs as markers to discriminate progeny from interspecies crosses. Developments in electrophoresis technology which have led to the introduction of pulse field systems is having a major impact on fungal genetics. Fortuitously, the chromosomes of most fungi fall in the size range of DNA molecules that can be separated by this method. Already the molecular karyotype of more than 40 species has been described including some of the basidiomycetes, *Agaricus bisporus*, (Horgen, personal communication), *Coprinus cinereus* (Pukkila, 1990) and *Phanaerochaete chrysosporium* (Gaskell *et al.*, 1991). The potential of this technique in the analysis of interspecies hybrids is clearly considerable. This technique identifies another useful application for protoplasts in that, so far, these naked cells have proved to be the most useful source of undamaged DNA.

Possibly the most powerful tool in hybrid analysis may be molecular karyotyping based on pulse field gel electrophoresis.

A limited survey of *Pleurotus* species (Table 3) revealed that several strains including *P. columbinus*, *P. cystidiosus*, *P. florida* and *P. pulmonaris* had a genome size in the range 17-20Mb. Four isolates of *P. sajor-caju* showed a wide variation in genome size suggesting that this group might include misidentified strains. A few strains had genomes greater that 25Mb and included small chromosomes which might be variable components of the genome.

TABLE 3. Chromosome numbers and estimated sizes for six species of *Pleurotus* as determined by CHEF electrophorersis.

SPECIES / STRAIN	BAND NUMBER AND ESTIMATED SIZE (Mb)										
	1	2	3	4	5	6	7	8	9	10	SIZE (Mb)
P. columbinus											
34-1	3.8	3.4	3.05	2.7	2.3	1.8					17.05
P. cystidiosus											
36-1	4.5	3.9	3.4	3	2.7	2.5					20.00
P. florida											
31-1	4.2	3.65	3.4	2.95	2.05	1.7	1.1				19.05
32-2	4.2	3.65	3.4	2.95	2.05	1.7	1.1				19.05
P. ostreatus											
7-4	>6	4.19	3.8	3.4	3.25	3	2.75	2.6	2.49	2.29	>39.49
P. pulmonaris											
33-1	5.5	3.9	3.2	2.7	2.3						17.60

TABLE 3 continued.

SPECIES / STRAIN	BAND NUMBER AND ESTIMATED SIZE (Mb)										
	1	2	3	4	5	6	7	8	9	10	SIZE (Mb)
P. sajor-caju											
32-1	5	4.05	3.1	2.9	2.75	2.5					20.30
32-2	>6	4.75*	4.19	3.5*	3.2	2.75	2.55				>34.99
32-3	>6	4	3	2.8	2.4						>18.20
32-4	4.6	4.1	3.5	3.2	2.9	2.7	2.5	2.4			25.90
P. sapidus											
35-1	>5	4.8	4.4	4	3.8	3.35	>1.7	1.7			>28.75

Possible doublet as estimated from fluroresence intensity.
(Fox, H.M. & Peberdy, J.F., unpublished results).

6. CONCLUSIONS

The usefulness of protoplasts in several areas of fungal biology is clearly established. The major area of interest has focused on their application as tools for genetic manipulation. In the basidiomycetes the most exciting aspect of this endeavour concerns the generation of novel interspecies hybrids. It is too early to consider whether a wide range of such hybrids can be produced because of the complex incompatibility mechanisms that operate and control the development of the all important fruit body. Furthermore, it is of great importance that true speciation is resolved in several genera before claims for interspecies hybridisation can be claimed. However, in the future basidiomycete fungi might feature in other useful roles including that of providing metabolites which have biological properties valuable to mankind. Basidiomycetes feature alongside the more established imperfect fungi in screens for the detection of pharmacologically important drugs. The improvement of any useful strain could involve protoplast fusion alleviating a need to establish conventional sexual crossing.

REFERENCES

ABE, M., UMETSU, H., NAKAI, T. & SASAGE, D. (1982). Regeneration and fusion of mycelial protoplasts of *Tricholoma matsutake*. *Agricultural and Biological Chemistry* **46**, 1955-1957.

ABE, M., NAKAI, T., UMETSU, H. & SASAGE, D. (1984). Regeneration of mycelial protoplasts from *Lyophyllum shimeji*. *Agricultural and Biological Chemistry* **48**, 1635-1636.

AKAMATSU, K., KAMADA, T. & TAKEMARU, T. (1983). Release and regeneration of protoplasts from the oidia of *Coprinus cinereus*. *Transactions of the Mycological Society of Japan* **24**, 173-184.

ANDERSON, J.B. & CENDESE, R. (1984). Extranuclear chloroamphenicol resistance mutations in the basidiomycete *Sistotrema brinkmannii*. *Experimental Mycology* **8**, 256-260.

ANDERSON, J.B., DETSCHE, D.M., HERR, F.B. & HORGEN, P.A. (1984). Breeding relationships among several species of *Agaricus*. *Canadian Journal of Botany* **62**, 1884-1889.

ANNE, J. & PEBERDY, J.F. (1975). Conditions for induced fusion of fungal protoplasts in polyethylene glycol solutions. *Archives of Microbiology* **105**, 201-205.

BARRETT, V., LEMKE, P.A. & DIXON, R.K. (1989). Protoplast formation from selected species of ectomycorrhizal fungi. *Applied and Environmental Microbiology* **30**, 381-387.

BURROWS, D.M., ELLIOTT, T.J. & CASSELTON, L.A. (1990). DNA mediated transformation of the secondarily homothallic basidiomycete *Coprinus bilanatus*. *Current Genetics* **17**, 175-177.

CASTLE, A.J., HORGEN, P.A. & ANDERSON, J.B. (1987). Restriction fragment length polymorphisms in the mushrooms *Agaricus brunnescens* and *Agaricus bitorquis*. *Applied and Environmental Microbiology* **53**, 816-822.

CHANG, S.T., LI, G.S.F. & PEBERDY, J.F. (1985). Isolation of protoplasts from edible fungi. *MIRCEN Journal of Applied Microbiology and Biotechnology* **1**, 185-194.

CHOI, S-H., KIM, B-K., KIM, H-W., KWAK, J-H., CHO, E-C., KIM, Y-C., YOO, Y-B. & PARK, Y-H. (1987). Studies on protoplast formation and regeneration of *Gandoderma lucidum*. *Archives of Pharmaceutical Research* **10**, 158-164.

DAVIS, B. (1985). Factors influencing protoplast isolation. In *Fungal Protoplasts. Applications in Biochemistry and Genetics*. pp. 45-71. Edited by J.F. Peberdy & L. Ferenczy. New York: Marcel Dekker, Inc.

DE VRIES, O.M.H. & WESSELS, J.G.H. (1972). Release of protoplasts from *Schizophyllum commune* by a lytic enzyme preparation from *Trichoderma viride*. *Journal of General Microbiology* **73**, 13-22 .

DE VRIES, O.M.H. & WESSELS, J.G.H. (1973). Effectiveness of a lytic enzyme preparation from *Trichoderma viride* in releasing spheroplasts from fungi, particularly basidiomycetes. *Antonie van Leeuwenhoek* **39**, 397-400.

DE VRIES, O.M.H. & WESSELS, J.G.H. (1975). Chemical analysis of cell wall regeneration and reversion of protoplasts from *Schizophyllum commune*. *Archives of Microbiology* **102**, 209-218.

EMERSON, S. & EMERSON, M.R. (1958). Production, reproduction and reversion of protoplast-like structures in the osmotic strain of *Neurospora crassa*. *Proceedings of the National Academy of Sciences of the United States of America* **44**, 668-671.

FERENCZY, L., KEVEI, F., SZEGEDI, M., FRANKO, A. & ROJIK, I.

(1976). Factors affecting high frequency fungal protoplast fusion. *Experientia* **32**, 1156-1158.

GASKELL, J., DIEPERINK, E. & CULLEN, D. (1991). Genomic organization of lignin peroxidase genes of *Phanerochaete chrysosporium. Nucleic Acids Research* **19**, 599-603.

GO, S-J., SHIN, G-C. & YOO, Y-B. (1985). Protoplast formation, regeneration and reversion in *Pleurotus ostreatus* and *P. sajor-caju. Korean Journal of Mycology* **13**, 169-177.

GOLD, M.H, CHENG, T.M. & ALIC, M. (1983). Formation, fusion, and regeneration of protoplasts from wild-type and auxotrophic strains of the white-rot basidiomycete *Phanerochaete chrysosporium. Applied and Environmental Microbiology* **46**, 260-263.

HAMYLN, P.F., BRADSHAW, R.E., MELLON, F.M., SANTIAGO, C.M., WILSON, J.M. & PEBERDY, J.F. (1981). Efficient protoplast isolation from fungi using commercial enzymes. *Enzyme and Microbial Technology* **3**, 321-325.

HASHIBA, T. & YAMADA, M. (1982). Formation and purification of protoplasts from *Rhizoctonia solani. Phytopathology* **72**, 849-852.

HERNANDEZ, E.S., MENDOZA, C.G. & NOVAES-LEDIEU, M. (1990). Chemical characterization of the hyphal walls of the basidiomycete *Armillaria mella. Experimental Mycology* **14**, 178-183.

HOCART, M.J., LUCAS, J.A. & PEBERDY, J.F. (1987). Production and regeneration of protoplasts from *Pseudocercosporella herpotrichoides* (Fron) Deighton. *Journal of Phytopathology* **119**, 193-205.

HOCART, M.J. & PEBERDY, J.F. (1989). Protoplast technology and strain selection. In *Biotechnology of Fungi for Improving Plant Growth.* pp. 235-258. Edited by J.M. Whipps & R.D. Lumsden. Cambridge: Cambridge University Press.

HOMOLKA, L., VYSKOCIL, P. & PILAT, P. (1988). Use of protoplasts in the improvement of filamentous fungi. I. Mutagenesis of protoplasts of *Oudemansiella mucida. Applied Microbiology and Biotechnology* **28**, 166-169.

HORGEN, P.A. & ANDERSON, J.B. (1989). Biotechnical advances in mushroom science. *Mushroom Science* **12**, 63-73.

IJIMA & YANAGI, S.O. (1986). A method for the high yield preparation of and high frequency regeneration of basidiomycete, *Pleurotus ostreatus* ("Hiratake") protoplasts using sulfite pulp waste components. *Agricultural and Biological Chemistry* **50**, 1855-1861.

KAWASUMI, T., KIUCHI, N., FUTATSUGI, Y., OHBA, K. & YANAGI, S.O. (1987). High yield preparation of *Lentinus edodes* ("Shiitake") protoplasts with regeneration capacity and mating type stability. *Agricultural and Biological Chemistry* **51**, 1649-1656.

KELKAR, H.S., SHANKAR, V. & DESHPANDE, V. (1990). Rapid isolation and regeneration of *Sclerotium folfsi* protoplasts and their potential for starch hydrolysis. *Enzyme and Microbial Technology* **12**, 510-515.

KIGUCHI, T. & YANAGI, S.O. (1985). Intraspecific heterokaryon and fruit body formation in *Coprinus macrorhizus* by protoplast fusion of auxotrophic mutants. *Applied Microbiology and Biotechnology* **22**, 121-4127.

KITAMOTO, Y., KONO R., TOKIMOTO, K., MORI, N. & ICHIKAWA, Y. (1984). Production of lytic enzymes against cell walls of basidiomycetes from *Trichoderma harzianum. Transactions of the Mycological Society of Japan* **26**, 69-79.

KITAMOTO, Y., KAGAWA, I., NAGAO, N., NAKAMATA, M. & ICHIKAWA, Y. (1987). High productivity protoplasting and reversion of protoplasts in *Coprinus cinereus* with a single preparation of lytic enzyme from *Trichoderma harzianum. Transactions of the Mycological Society of Japan* **28**, 217-228.

KITAMOTO, Y., MORI, N., YAMAMOTO, M., OHIWA, T. & KHIKAWA, Y. (1988). A simple method for protoplast formation and improvement of protoplast regeneration from various fungi using an enzyme from *Trichoderma harzianum. Applied Microbiology and Biotechnology* **28**, 445-450.

KOGA, D., SUESHIGE, N., OKIKONO, K., UTSUMI, T., TANAKA, S., YAMADA, Y. & IDE, A. (1988). Efficiency of chitinolytic enzymes in the formation of *Tricholoma matsutake* protoplasts. *Agricultural and Biological Chemistry* **52**, 2091-2093.

KROPP, B.R. & FORTIN, J.A. (1985). Formation and regeneration of protoplasts from the ectomycorrhizal basidiomycete *Laccaria bicolor. Canadian Journal of Botany* **64**, 1224-1226.

KUWABARDA, H., MAGAE, Y., KASHIWAGI, Y., OKADA, G. & SASAKI, T. (1989). Characterization of enzyme productivity of protoplast regenerants from the cellulase-producing fungus *Robillarda* Y-20. *Enzyme and Microbial Technology* **11**, 696-699.

LAU, W.C., DHILLON, E.K.S. & CHANG, S.T. (1985). Isolation and reversion of protoplasts of *Pleurotus sajor-caju*. *MIRCEN Journal of Applied Microbiology and Biotechnology* **1**, 345-353.

LEIN, J. (1986). The Panlabs strain improvement programme. In *Overproduction of Microbial Metabolites*. pp. 105-140. Edited by Z. Vanek & Z. Hostalek. Boston: Butterworths.

MAGAE, Y., KAKIMOTO, Y., KASHIWAGI, Y. & SASAKI, T. (1985). Fruiting body formation from regenerated mycelium of *Pleurotus ostreatus* protoplasts. *Applied and Environmental Microbiology* **9**, 441-442.

MELLON, F.M., LITTLE, P.F.R. & CASSELTON, L.A. (1987). Gene cloning and transformation in the basidiomycete fungus *Coprinus cinereus*: Isolation and expression of the isocitrate lyase gene (*acu-7*). *Molecular and General Genetics* **210**, 352-357.

MENDOZA, C.G., AVELLAN, M.A., SANCHEZ, E., & NOVAES-LEDIEU, M. (1987). Differentiation and wall chemistry of *Agaricus bisporus* vegetative and aggregated mycelia. *Archives of Microbiology* **148**, 68-71.

MENDOZA, C.G., CABO, A.P., GONZALEZ, M.L.S. & NOVAES-LEDIEU, M, (1991). Morphological and ultrastructural studies on protoplast production and reversion of the higher basidiomycete *Agaricus bosporus*. *Current Microbiology* **22**, 191-194.

MOORE, D. (1975). Production of *Coprinus* protoplasts by use of chitinase or helicase. *Transactions of the British Mycological Society* **65**, 134-136.

MORINAGA, T., KIKUCHI, M. & NOMI, R. (1985). Formation and regeneration of protoplasts in *Coprinus pellucidus* and *Coprinus cinereus*. *Agricultural and Biological Chemistry* **49**, 523-524.

MUKHERJEE, M. & SENGUPTA, S. (1986). Mutagenesis of protoplasts and regeneration of mycelium in the mushroom *Volvariella volvacea*. *Applied and Environmental Microbiology* **52**, 1412-1414.

MUKHERJEE, M. & SENGUPTA, S. (1988). Isolation and regeneration of protoplasts of *Termitomyces clypeatus*. *Canadian Journal of Microbiology* **34**, 1330-1332.

MUÑOZ-RIVAS, A., SPECHT, C.A., DRUMMOND, B.J., FROELIGER, E. & NOVOTNY, C. (1986). Transformation of the basidiomycete, *Schizophyllum commune*. *Molecular and General Genetics* **295**, 103-106.

NOEL, T. & LABARERE, J. (1989). Isolation and reversion of protoplasts from homokaryotic mycelium of *Agrocybe aegerita*. *Mushroom Science* **12**, 175-185.

OHMASA, M. (1986). Intraspecific protoplast fusion of *Pleurotus ostreatus* using auxotrophic mutants. *Japanese Journal of Breeding* **36**, 429-433.

OHMASA, M., ABE, Y., FURUKAWA, H., TANIGUCHI, M. & NEDA, H. (1987). Preparation and culture of protoplasts of some Japanese cultivated mushrooms. *Bulletin of Forestry & Forest Products Research Institute* **343**, 155-170.

PARK, Y-D., YOO, Y-B., SHIN, P-G., YOU, C-H., CHA, D-Y., PARK, Y-H. & LEE, J-S. (1988). Interspecific protoplast fusion of *Gandoderma applanatum* and *Gandoderma lucidum* and fruit body formation of the fusants. *Korean Journal of Mycology* **16**, 79-86.

PEBERDY, J.F. (1979). Fungal protoplasts: isolation, reversion, and, fusion. *Annual Review of Microbiology* **33**, 21-29.

PEBERDY, J.F. (1989). Fungi without coats - protoplasts as tools for mycological research. *Mycological Research* **93**, 1-20.

PEBERDY, J.F. (1990). Fungal Cell Walls - A review. In *Biochemistry of Cell Walls and Membranes in Fungi* pp. 5-30. Edited by P.J. Kuhn, A.P.J. Trinci. M.J. Jung, M.W. Goosey & L.G. Copping. Berlin: Springer Verlag.

PEBERDY, J.F. (1991). Fungal Protoplasts. In *More Gene Manipulations in Fungi*. Edited by J.W. Bennett & L.A. Lasure. San Diego: Academic Press. In press.

PEBERDY, J.F. & HOCART, M.J. (1987). Protoplasts as a tool in the genetics of plant pathogenic fungi In *Genetics and Plant Pathogenesis*. pp. 127-142. Edited by P.R. Day & G.J. Jellis. Oxford, UK: Blackwell Scientific Publications.

PUKKILA, P.J. (1990). Methods of genetic manipulation in *Coprinus cinereus*. In *World-wide Progress of Mushroom Technology*: Abstracts of the satellite symposium of the IUMS congress: Bacteriology and Mycology, Osaka, Japan.

ROYSE, D.J. & MAY, B. (1982a). Use of isozyme variation to identify geotypic classes of *Agaricus brunnescens*. *Mycologia* **74**, 93-102.

ROYSE, D.J. & MAY, B. (1982b). Genetic relatedness and its application in selective breeding of *Agaricus brunnescens*. *Mycologia* **74**, 569-575.

ROYSE, D.J., SPEAR, M.C. & MAY, B. (1983a). Cell line authentication and genetic relatedness of lines of the Shiitake mushroom, *Lentinus edodes. Journal of General and Applied Microbiology* **29**, 205-216.

ROYSE, D.J., SPEAR, M.C. & MAY, B. (1983b). Single and joint segregation of marker loci in the Shiitake mushroom, *Lentinus edodes. Journal of General and Applied Microbiology* **29**, 217-222.

ROYSE, D.J., JORDAN, M.H., ANTOUN, G.G. & MAY, B. (1987). Confirmation of intraspecific and single and joint segregation of biochemical loci of *Volvariella volvacea. Experimental Mycology* **11**, 11-18.

SANTIAGO, C.M. (1982a). Protoplast isolation in the common tropical mushroom using microbial enzyme. *Kalikasan Philippine Journal of Biology* **11**, 365-371.

SANTIAGO, C.M. (1982b). Production of *Volvariella* protoplasts by use of *Trichoderma* enzyme. *Mushroom Newsletter for the Tropics* **3**, 3-6.

SHIN, G.C., YEO, U.H., YOO, Y.B. & PARK, Y.H. (1986). Some factors affecting the protoplast formation and regeneration from the mycelium of *Gandoderma lucidum* (Fr.) Karsten. *Research Reports in Agricultural Science and Technology* **13**, 185-192.

SONNENBERG, A.S.M. & WESSELS, J.G.H. (1987). Heterokaryon formation in basidiomycetes by electrofusion of protoplasts. *Theoretical and Applied Genetics* **74**, 654-658.

SONNENBERG, A.S., WESSELS, J.G.H. & VAN GRIENSVEN, L.J. (1988). An efficient protoplasting/regeneration system for *Agaricus bisporus* and *Agaricus bitorquis. Current Microbiology* **17**, 285-291.

STILLE, B. (1984). Release of protoplasts from *Volvariella bombycina* (Schaeff. Ex. Fr.) singer, a method developed. *Mushroom Newsletter for the Tropics* **4**, 15.

STRUNK, C. (1965). Uber entstehung und reversion enzymatisch erzeugter Protoplasten von *Polystictus versicolor. Biologische Rundschau* **3**, 242-244.

THOMAS, K.R. & DAVIS, B. (1980). The effect of calcium on protoplast release from species of *Aspergillus. Microbios* **28**, 69-80.

TOYODA, H., HIRAI, T., SUMIYA, H., KAWAKAMI, Y., SAKAMOTO, M. & USHIYAM R. (1984). Preparation and cell wall regeneration of protoplasts of *Lentinus edodes. Memoirs of the Faculty of Agriculture Kinki University* **17**, 121-130.

TOYOMASU, T., MATSUMOTO, T. & MORI, K. (1986). Interspecific protoplast fusion between *Pleurotus ostreatus* and *Pleurotus salmoneo-*

stramineus. Agricultural and Biological Chemistry **50**, 223-225.

TOYOMASU, T. & MORI, K. (1987a). Intra- and interspecific protoplast fusion between some *Pleurotus* species. *Agricultural and Biological Chemistry* **51**, 935-937.

TOYOMASU, T. & MORI, K. (1987b). Fruit body formation of the fusion products obtained on interspecific protoplast fusion between *Pleurotus* species. *Agricultural and Biological Chemistry* **51**, 2037-2040.

ULLRICH, R.C., NOVOTNY, C.P., SPECHT, C.A., FROELIGER, E.A. & MUÑOZ-RIVAS, A.M. (1985). Transforming basidiomycetes. In *Molecular Genetics of Filamentous Fungi*. pp. 39-57. Edited by W. Timberlake. New York: A.R. Liss Inc.

UM, S.D., CHAE, Y.A., YOO, Y.B., YOU, C.H. & CHA, D.Y. (1988). Protoplast isolation and reversion from *Gandoderma lucidum* and *Gandoderma* sp. *Korean Journal of Mycology* **16**, 21-25.

USHIYAMA, R. & NAKAI, Y. (1977). Protoplasts of Shiitake, *Lentinus edodes* (BERK.) SING. *Report of the Tottori Mycological Institute (Japan)* **15**, 1-5.

VAN DER VALK, H.C.P.M. & WESSELS , J.G.H. (1976). Ultrastructure and localization of wall polymers during regeneration and reversion of protoplasts of *Schizophyllum commune* protoplasts. *Protoplasma* **90**, 65-87.

WAKABAYASHI, S., MAGAE, Y., KASHIWAGI, Y. & SASAKI, T. (1985). Formation of giant protoplasts from protoplasts of *Pleurotus cornucopiae* by the cell wall lytic enzyme. *Applied Microbiology and Biotechnology* **21**, 328-330.

WESSELS, J.G.H. & SIETSMA , J.H. (1979). Wall structure and growth in *Schizophyllum commune*. In *Fungal Walls and Hyphal Growth* pp. 27-48. Edited by J.H. Burnett & A.P.J. Trinci. Cambridge: Cambridge University Press.

WOO, H.S. & YOON, Y. (1985). Formation and regeneration of protoplasts in *Lentinus edodes. Mushroom Newsletter for the Tropics* **5**, 4-10.

WU, M.M.J., CASSIDY, J.R. & PUKKILA, P.J. (1983). Polymorphisms in DNA of *Coprinus cinereus. Current Genetics* **7**, 385-392.

YAMADA, O., MAGAE, Y., KASHIWAGI, Y., KAKIMOTO, Y. & SASAKI, T. (1983). Preparation and regeneration of mycelial protoplasts of *Collybia velutipes* and *Pleurotus ostreatus. European Journal of Applied Microbiology and Biotechnology* **17**, 298-300.

YANAGI, S.O. & TAKEBE, I. (1984). An efficient method for the isolation of mycelial protoplast from *Coprinus macrorhizus* and other basidiomycetes. *Applied Microbiology and Biotechnology* **19**, 58-60.

YANAGI, S.O., MONMA, M., KAWASUMI, T., HINO, A., KITO, M. & TAKEBE, I. (1985). Conditions for isolation of and colony formation by mycelial protoplasts of *Coprinus macrorhizus*. *Agricultural and Biological Chemistry* **49**, 171-179.

YEA, U-H., YOO, Y-H., PARK, Y-H. & SHIN, G-C. (1988). Isolation of protoplasts from *Flammulina velutipes*. *Korean Journal of Mycology* **16**, 70-78.

YOO, Y-B. (1989). Fusion between protoplasts of *Gandoderma applanatum* and oidia of *Lyophyllum ulmarium*. *Korean Journal of Mycology* **17**, 197-201.

YOO, Y-B., BYUN, M-O., GO, S-J., YOU, C-H., PARK, Y-H. & PEBERDY J.F. (1984). Characteristics of fusion products between *Pleurotus ostreatus* and *Pleurotus florida* following interspecific protoplast fusion. *Korean Journal of Mycology* **12**, 164-169.

YOO, Y-B., PEBERDY, J.F. & YOU, C-H. (1985). Studies on protoplast isolation from edible fungi. *Korean Journal of Mycology* **13**, 1-10.

YOU, C-H., YOO, Y-B. & PARK, Y-H. (1988). Studies on protoplast formation and reversion of *Pleurotus sapidus* Kalchbur. *Korean Journal of Mycology* **16**, 214-219.

YU, M.Y. & CHANG, S.T. (1987). Effects of osmotic stabilizers in the activities of mycolytic enzymes used in fungal protoplast liberation. *MIRCEN Journal of Applied Microbiology and Biotechnology* **3**, 161-167.

CHAPTER 8

GENE TRANSFER IN EDIBLE FUNGI USING PROTOPLASTS

Young Bok Yoo and Dong Yeul Cha

Applied Mycology & Mushroom Division, Agricultural Science Institute, Suweon 440-707, Republic of Korea.

1. INTRODUCTION

The objective of mushroom breeding is strain improvement by conventional and non-conventional means. The conventional method is one of mankind's most ancient applications of biotechnology and still has much to offer. Some methods of non-conventional breeding have now been established in fungi.

Gene transfer using protoplasts has been developed to break down the barrier of gene exchange imposed by conventional breeding systems. A few distinct methods of gene transfer in fungi are protoplast fusion, uptake of foreign genetic materials and transformation. The development of a gene transfer system in edible fungi would be useful in understanding the mechanism of genetic recombination and strain improvement. For additional information on gene transfer using protoplast techniques, the reader is referred to several reviews that have already appeared in the literature (Ferenczy, 1984; Wu, 1987; Fincham, 1989; Peberdy, 1989). This review will discuss some fundamental and practical aspects associated with recent advances in protoplast technology involving higher fungi.

2. PROTOPLAST ISOLATION AND REGENERATION

Isolated protoplasts are naked cells that have had their cell wall removed either by mechanical action or by enzyme digestion. Cell wall degrading enzymes are toxic to varying degrees and might unfavourably affect the physiology of the cells. However, the routine formation of protoplasts has become possible by the availability of a number of commercial lytic enzymes. As a result of wall removal, the only barrier that exists between the cell protoplasm and the external environment is the plasma membrane.

Protoplast technology depends on the availability of protoplasts in large numbers. Recently published and unpublished data on protoplast isolation from Basidiomycotina are described in Table 1. Formation of protoplasts from the mycelia of edible fungi was influenced by some major factors such as lytic enzyme, the osmotic stabilizer and the physiological status of the organism. For isolation of protoplasts, the choice of lytic enzyme for digesting the cell walls is more crucial. Novozym 234 (Novo), Cellulase CP (Sturge), Cellulase onozuka R-10 (Yakult), Chitinase (Sigma), β-Glucanase (BDH) and Snail enzyme were used to compare the effect on the release of protoplasts from edible fungi. Novozym 234 was the most effective enzyme for the fungi. The highest yield of protoplasts from several species was obtained by using a combined lytic enzyme system containing Novozym 234, β-glucanase and β-glucuronidase. Analysis of the cell wall of filamentous fungi revealed that the major components were chitin, β-glucans, α-glucan, glyco-proteins and chitosan (Farkas, 1985; Peberdy, 1990). The Novozym 234 was found to contain considerable activities of chitinase, β-D-glucanase and protease compared with other commercial enzymes (Hamlyn *et al.*, 1981; Peberdy, 1985). Most mycolytic enzymes currently in use are derived from microorganisms except snail enzyme. The gastric juice from the snail *Helix pomatia* is available under several trade names such as β-glucuronidase, Glusulase and Helicase.

Once the cell wall has been digested away, the isolated protoplast is subject to osmotic stress. In naked protoplasts, the wall pressure must be replaced by osmotic pressure in the isolation mixture. Inorganic and organic osmotic stabilizers used were KCl, $MgSO_4$, NaCl, NH_4Cl, $(NH_4)_2SO_4$, glucose, mannitol, sorbitol and sucrose. Among the osmotic stabilizers, Sucrose 0.6M was the most effective for the release of protoplasts from 2-6 day old mycelia on permeable cellophane membrane overlaying complete

TABLE 1. Protoplast isolation from mycelia of edible fungi.

Species	Enzyme	Osmoticum	Mycelial age (day)	Protoplast (x $10^6 ml^{-1}$)	Reference[1]
Agaricus bisporus	Novozym 234 β-glucanase Snail enzyme	$MgSO_4$	4	0.02	Yoo *et al.* (1985)
Coriolus versicolor	Novozym 234 Cellulase onozuka	Sucrose	2.5	30.80	Bok *et al.* (1990)
Elfvingia applanata	Novozym 234 Cellulase onozuka β-glucuronidase	Sucrose	3	50.40	Park *et al.* (1987b)
Flammulina velutipes	Novozym 234 Cellulase CP	Sucrose	5	5.20	Yea *et al.* (1988)
Ganoderma lucidum	Novozym 234	Sucrose	3	4.02	Shin *et al.* (1986)
	Novozym 234 β-glucuronidase	Sucrose	4	5.96	Choi *et al.* (1987)
Hypsizigus marmoreus	Novozym 234 β-glucanase β-glucuronidase	Sucrose	6	26.15	Yoo *et al.* (1987c)
Lentinus edodes	Novozym 234 β-glucanase Snail enzyme	$MgSO_4$	5	10.00	Yoo *et al.* (1985)
Pleurotus cornucopiae	Novozym 234 β-glucanase β-glucuronidase	Sucrose	4	11.60	Lee *et al.* (1986a)
Pleurotus florida	Novozym 234 Cellulase CP	Sucrose	3	4.88	
Pleurotus ostreatus	Novozym 234 β-glucanase β-glucuronidase	Sucrose	4	37.48	
	Novozym 234	Sucrose	3	26.40	Go *et al.* (1985)

TABLE 1 continued

Species	Enzyme	Osmoticum	Mycelial age (day)	Protoplast (x $10^6 ml^{-1}$)	Reference[1]
Pleurotus sajor-caju	Novozym 234 β-glucanase β-glucuronidase	Sucrose	4	7.38	
Pleurotus salmoneo-stramineus	Novozym 234 β-glucanase β-glucuronidase	Sucrose	4	4.92	Yoo *et al.* (1989a)
Pleurotus sapidus	Novozym 234 β-glucanase β-glucuronidase	Sucrose	4	31.70	You *et al.* (1988b)
Pleurotus spodoleucus	Novozym 234	Sucrose	3	3.35	Yoo *et al.* (1987a)
Volvariella volvacea	Novozym 234 β-glucanase Snail enzyme	$MgSO_4$	2	0.07	Yoo *et al.* (1985)

[1] Y. B. Yoo unpublished data

agar medium. KCl, $MgSO_4$, and NaCl interfered with the enzyme activity for protoplast isolation from higher fungi. Yu and Chang (1987) examined the activity of Novozym 234 in the presence of various osmotic stabilizers. β-glucanase was inhibited more by KCl and NaCl than by sucrose. Incidentally, inorganic stabilizers such as KCl, $MgSO_4$, and NH_4Cl stimulated the best protoplast yield in Ascomycotina and Deuteromycotina (Peberdy *et al.*, 1976; Davis, 1985). The pH and the buffer used in the incubation medium may also affect protoplast yield. Phosphate buffer with 0.6M sucrose was more effective than Na-maleate buffer for the release of protoplast from *Pleurotus cornucopiae*. However, 0.6M sucrose osmotic stabilizer without pH adjustment was the most effective (Lee *et al.*, 1986a). On the other hand, Na-maleate buffer with 0.6M $MgSO_4$ was more effective than phosphate buffer for releasing protoplasts (De Vries & Wessels, 1972; Yoo *et al.*, 1985). In *P. ostreatus*, protoplast production from liquid-cultured mycelia in erlenmeyer

TABLE 2. Frequency of monokaryotic strains recovered from protoplast regeneration colonies of edible fungi (Modified from Yoo *et al.* 1987b).

Species	No. of colonies examined	No. of monokaryon	% of monokaryon	% of protoplast reversion
Flammulina velutipes	500	76	15.20	0.47 - 1.32
Ganoderma lucidum	742	85	11.46	0.47 - 1.47
Hypsizigus marmoreus	313	14	4.47	0.0002 - 0.23
Pleurotus cornucopiae	100	18	18.00	0.01 - 1.29
Pleurotus ostreatus	400	22	5.50	0.02 - 0.75
Pleurotus salmoneo -stramineus	204	2	0.98	0.44 - 0.68
Pleurotus sapidus	100	0	0	0.18 - 2.02
Pleurotus spodoleucus	402	196	48.76	0.01 - 0.28

flasks was less than that from surface-grown mycelia on permeable cellophane membranes (Yoo *et al.,* 1985). The low yields of protoplast formation in the liquid media were similar to those obtained with other filamentous fungi (Ferenczy *et al.*, 1975).

Fungal protoplasts are usually cultured in either liquid or semisolid media. Patterns of regeneration and reversion of *Pleurotus ostreatus* and *P. cornucopiae* after incubation for 1-3 days on solid minimal medium (MM;

Raper *et al.*, 1972) showed formation of a yeast-like cell chain and direct development of 1-4 germ tubes from a spherical protoplast. The complete medium (CM; Raper *et al.*, 1972) stabilized with 0.6M sucrose was the most effective for regeneration of protoplasts in higher fungi and there may be a correlation between osmotic stabilizer and metabolizable nutrient source in regeneration media. Lee *et al.* (1986b) reported that there was a correlation between the nutrient source for protoplast regeneration and that for hyphal growth. The rates of protoplast regeneration and hyphal growth of *P. cornucopiae* were stimulated by various amino acids and nucleic acid components as well as some vitamins. When reversion colonies of protoplasts from eight dikaryotic species were transferred to CM or MM they showed clampless monokaryotic colonies except in the case of one species (Table 2). More than four types of colony morphology were detected among neohaplonts from protoplast reversion in *Hypsizigus marmoreus*, *P. cornucopiae*, and *P. spodoleucus*; *Flammulina velutipes*, *Ganoderma lucidum* and *P. ostreatus* showed one or two types. Growth rates of neohaplonts were slow compared with those of dikaryotic colonies. Monokaryotic strains have been obtained by micro-surgery (Harder, 1927; Aschan, 1952), poisonous chemicals (Miles & Raper, 1956; Parag, 1961; Kerruish & Dacosta, 1963; Takemaru & Kamada, 1971), submerged culture (Ginterova, 1973) and protoplast regeneration (Wessels *et al.*, 1976). Protoplasts from a dikaryon of *Schizophyllum commune* reverted into monokaryotic mycelia at a frequency of 20-40%. These monokaryotic strains could be very useful for studying the genetic control of sexuality in higher fungi.

3. PROTOPLAST FUSION

3.1. Intraspecies and Interspecies Protoplast Fusion

3.1.1. Protoplast fusion and incompatibility. When protoplasts are isolated, the only barrier between the cytoplasm and the external environment is the plasma membrane. The lack of the cell wall allows the plasma membrane of two or more protoplasts to come into intimate contact, something which is not possible under normal circumstances. The sequence of fusion events requires approach, adhesion, lipid-lipid interaction, cytoplasmic bridge and fusion.

Auxotrophic mutants were induced following the U.V. irradiation of protoplasts, basidiospores and mycelial fragments of *Pleurotus* spp. The following 64 mutants were identified by means of a modified Holliday solution : 26 requiring amino acids, 14 requiring vitamins, 7 requiring nucleic acid bases and related compounds, 11 requiring amino acids-vitamins, 3 requiring amino acids-nucleic acid bases and related compounds, and 3 requiring vitamins-nucleic acid bases and related compounds (After Yoo *et al.*, 1988a). Twelve spontaneous auxotrophs were also obtained by protoplast regeneration of *P. salmoneostramineus*, and all were amino acid-requiring strains (After Yoo *et al.*, 1989a). Intraspecies and interspecies fusion products were obtained by polyethylene glycol-induced fusion of protoplasts from auxotrophic mutants of *Pleurotus* (Table 3). The fusants produced colonies with dense mycelium after 10-20 days culture on hypertonic MM plates. The sectors showed normal vegetative morphology but the colonies of irregular shape varied in growth rate. Intra- and interspecific protoplast fusants between compatible strains in pairings of *P. ostreatus* + *P. ostreatus*, *P. florida* + *P. ostreatus*, *P. florida* + *P. sajor-caju*, and *P. ostreatus* + *P. sajor-caju* formed true clamp connections on CM, and produced fruit bodies rapidly andabundantly in bottles containing poplar sawdust plus rice bran medium. The conventional method of hyphal anastomosis has been used to force the production of intra- and interspecific heterokaryons between compatible strains in *Pleurotus*. On MM, the interspecific fusants between vegetatively incompatible strains in pairings of *P. cornucopiae* + *P. florida*, *P. cornucopiae* + *P. salmoneostramineus*, *P. florida* + *P. sajor-caju*, *P. florida* + *P. salmoneostramineus*, *P. ostreatus* + *P. sajor-caju*, *P. sajor-caju* + *P. salmoneostramineus*, and *P. sajor-caju* + *P. sapidus* developed very slowly. Mycelial colonies were very distorted and composed of thick-layered hyphae. When transferred to CM plates, the fusants were classified into allodiploids, stable heterokaryons and spontaneous segregants. Heterokaryons between incompatible strains grew less vigorously than compatible fusants. With the exception of a few heterokaryons, these interspecific fusants between incompatible pairs of strains did not form true clamp connections, and did not produce fruit bodies on complete agar medium or complete liquid medium. However, fruitbody production by clampless fusants was induced by light-dark cycles on a medium containing 570g poplar tree sawdust plus 20% rice bran in an 1000ml glass bottle. All of the primordia from fusants between incompatible strains showed clamp connections except the combinations of *P. cornucopiae* + *P. florida* and *P. florida* + *P.*

TABLE 3. Characteristics of fusion products of protoplasts in higher fungi.

Strain[1)]	Clamp[2)]		Fruiting		Genetic background	Reference[4)]
	Anastomosis	Protoplast fusion	Type	Clamp[3)]		
Pleurotus ostreatus-m +	+	+	*Pleurotus ostreatus*	+	heterokaryon	
Pleurotus ostreatus-m	-	-	non-fertile		heterokaryon	
Pleurotus cornucopiae-m +	-	+	intermediate	+	heterokaryon	Yoo (1991)
Pleurotus florida-m		-	intermediate	+	heterokaryon	
		-	primordia	-	heterokaryon	
		-	non-fertile		heterokaryon	
		-	non-fertile		nuclear hybrid	
Pleurotus cornucopiae-m +	-	+	*Pleurotus salmoneostramineus*	+	heterokaryon	
Pleurotus salmoneostramineus-d		-	non-fertile		heterokaryon	
Pleurotus florida-m + *Pleurotus ostreatus*-m	+	+	intermediate	+	heterokaryon	Yoo *et al.* (1984)
Pleurotus florida-m +	+	+	intermediate	+	heterokaryon	after Yoo *et al.* (1987d)
Pleurotus sajor-caju-m	-	-	intermediate	+	heterokaryon	
		-	primordia	-	heterokaryon	

TABLE 3. continued

Strain[1)]	Clamp[2)]		Fruiting		Genetic background	Reference[4)]
	Anastomosis	Protoplast fusion	Type	Clamp[3)]		
Pleurotus florida-m +	-	+	*Pleurotus salmoneostramineus*	+	heterokaryon	
Pleurotus salmoneostramineus-d		-	*Pleurotus florida*	-	heterokaryon	
		-	non-fertile		heterokaryon	
Pleurotus ostreatus-m +	+	+	*Pleurotus ostreatus*	+	heterokaryon	after Yoo *et*
Pleurotus sajor-caju-m	-	-	intermediate	+	heterokaryon	*al.* (1987d)
		-	primordia	-	heterokaryon	
Pleurotus sajor-caju-m +	-	+	*Pleurotus salmoneostramineus*	+	heterokaryon	
Pleurotus salmoneostramineus-d		-	non-fertile		heterokaryon	
Pleurotus sajor-caju-m +	+	+	intermediate	+	heterokaryon	
Pleurotus sapidus-d						
Elfvingia applanata-d +	-	+	*Elfvingia applanata*	+	heterokaryon	Park (1988)
Ganoderma lucidum-d		+	*Ganoderma lucidum*	+	heterokaryon	Park *et al.*
		+	intermediate	+	heterokaryon	(1988)
		-	non-fertile		heterokaryon	Um *et al.* (1988)

TABLE 3. continued

Strain[1]	Clamp[2]		Fruiting		Genetic background	Reference[4]
	Anastomosis	Protoplast fusion	Type	Clamp[3]		
Pleurotus ostreatus-m + *Elfvingia applanata*-d	-	+	*Pleurotus ostreatus*	+	synkaryon	after Yoo *et al.* (1989c)
		-	*Pleurotus ostreatus*	+	synkaryon	
		-	non-fertile		synkaryon	

1) m (monokaryon), d (dikaryon)
2), 3) + (Present clamp connection), - (Absent clamp connection)
4) Y. B. Yoo unpublished data

salmoneostramineus. None of these clampless primordia produced mature basidiocarps except one fusant of *P. florida* + *P. salmoneostramineus*. Certain hyphae of the clamped primordia formed on sawdust medium also formed clamp connections. When small pieces of stipe tissue from the primordium or basidiocarp were cultured on CM plates, mycelial colonies grew more vigorously than the original incompatible fusants and the hyphae of the colony produced clamp connections.

The major species of *Pleurotus* are all bifactorial heterothallic. Single-spore isolates from basidiocarps are homokaryotic and self-sterile (Vandendris, 1932, 1933; Terakawa, 1960; Eugenio & Anderson, 1968; Anderson *et al.*, 1973). Eger, 1974; Roxon & Jong, 1977; Hilber, 1982). However, homokaryotic fruiting has been reported in some species of *Pleurotus* including *P. eous* (Elliott, 1985), *P. flabellatus* (Samsudin & Graham, 1984), *P. florida* (Y. B. Yoo unpublished results), *P. ostreatus* (Eger, 1974) and *P. sajor-caju* (Go & Shin, 1986; Liang & Chang, 1989). Monokaryotic auxtotrophs of *Pleurotus* used in these experiments were self-sterile except *P. sajor-caju*. Six strains of *P. sajor-caju* formed primordium initials or primordia and one strain, ASI 2-45-lys, developed mature fruit bodies on sawdust medium. Dikaryotic auxotrophs of *P. salmoneostramineus* and *P. sapidus* produced mature fruit bodies. Interspecific hybrids have been obtained by hyphal anastomosis in crosses between homokaryotic strains of *P. abalonus* x *P. cystidiosus* (Hilber, 1982), *P. eryngii* x *P. ferulae* (Hilber, 1982), *P. eryngii* x *P. nebrodensis* (Hilber, 1982), *P. ferulae* x *P. nebrodensis* (Hilber, 1982), *P. florida* x *P. ostreatus* (Go *et al.*, 1981; Hilber, 1982, Yoo *et al.*, 1987d), *P. florida* x *P. pulmonarius* (Hilber, 1982), *P. florida* x *P. sajor-caju* (Yoo *et al.*, 1987d), *P. ostreatus* x *P. sajor-caju* (Yoo *et al.*, 1987d; May *et al.* 1988), *P. ostreatus* x *P. sapidus* (May *et al.*, 1988), and *P. sajor-caju* x *P. sapidus* (May *et al.*, 1988). Intra- and interspecific protoplast fusion in *Pleurotus* has been reported in several earlier papers (Toyomasu *et al.*, 1986; Yoo *et al.*, 1986; Yoo *et al.*, 1987d; Go *et al.*, 1989; Toyomasu & Mori, 1989). However, these authors failed to obtain clamped mature basidiocarps from clampless fusion products derived from vegetatively incompatible pairs of strains. Intraspecific heterokaryons between antagonistic morphological variants of *Volvariella volvacea* were obtained by protoplast fusion, and the fusants produced fruiting bodies (Santiago, 1981). The pattern of sexuality in *V. volvacea* is primary homothallism, in which incompatibility factor is absent. On the other hand, incompatible heterokaryons of *Coprinus macrorhizus* did not develop

TABLE 4. Segregation and recombination of genetic markers in progenies of interspecific hybrids of *Pleurotus* following protoplast fusion (Y. B. Yoo unpublished results).

Strain combination[1)]				Parentals		Recombinant[3)]		
A	B	Fusant	Compatibility[2)]	A	B	Pro.	Auxo.	% of recomb.
P. c 2-29-ade	+P. f 2-3-ura	P549	-	34	3	70	11	78.43
		P564		22	0	51	29	68.64
P. c 2-28-gln cit arg	+P. ss 2-63-leu arg gly	P709	-	0	0	230	10	100
		P715		0	0	236	4	100
P. c 2-39-cit pan	+P. ss 2-63-leu arg gly	P728	-	1	0	317	2	99.68
		P730		1	0	316	3	99.68
P. f 2-3-rib	+P. o 2-1-arg	P 3	+	129	0	191	0	59.68
		P 5		178	39	212	33	53.03
P. f 2-3-rib	+P. o 2-2-gly ser	P 34	+	168	0	105	1	38.68
P. f 2-4-rib	+P. o 2-1-arg	P 22	+	98	9	94	43	56.14
		P 25		46	0	62	34	67.60
P. f 2-4-rib	+P. o 2-2-gly ser	P 12	+	72	0	84	0	53.84
		P 15		227	0	240	11	52.51
P. f-P.o-arg rib	+P. o 2-13-pro orn	P 48	+	17	0	136	43	91.32
		P 49		2	0	96	130	99.12
		P 72		1	0	75	152	99.56

TABLE 4. continued.

Strain combination[1)]				Parentals		Recombinant[3)]		
A	B	Fusant	Compat-ibility[2)]	A	B	Pro.	Auxo.	% of recomb.
P. f 2-3-rib	+P. s 2-44-lys	P150	-	1	10	97	12	90.83
P. f 2-3-rib	+P. s 2-55-ane pn asn	P113	+	34	0	70	8	69.64
		P122		33	1	51	19	66.66
P. f 2-3- rib	+P. ss 2-63-leu arg gly	P793	-	1	0	707	12	99.86
		P795		1	0	544	1	99.81
P. o 2-1-arg	+P. s 2-45-lys	P179	-	1	0	209	110	99.68
P. o 2-1-arg	+P. s 2-53-rib ane	P207	-	0	1	226	93	99.68
P. o 2-2-gly ser	+P. s 2-45-lys	P171	-	12	2	303	15	99.37
P. o 2-2-gly ser	+P. s 2-47-orn ala	P166	-	1	13	273	33	95.62
		P168		0	2	273	29	99.34
P. o 2-2-gly ser	+P. s 2-49-pan	P152	-	12	3	241	32	94.79
		P154		4	2	270	44	98.12
P. o 2-2-gly ser	+P. s 2-55-ane pn asn	P157	+	0	0	104	16	100
P. s 2-49-pan	+P. ss 2-63-leu arg gly	P746	-	0	0	182	26	100
		P757		0	0	192	16	100

TABLE 4. continued.

Strain combination[1)]				Parentals		Recombinant[3)]		
A	B	Fusant	Compat-ibility[2)]	A	B	Pro.	Auxo.	% of recomb.
P. s 2-49-pan	+P. sd 2-56-gly arg cyn	P787	-	50	0	64	4	57.62
		P788		76	0	211	17	75.00

1) P. c (*P. cornucopiae*), P. f (*P. florida*), P. o (*P. ostreatus*), P. s (*P.sajor-caju*), P. sd (*P. sapidus*), P. ss (*P. salmoneostramineus*), ade (adenine), ala (alanine), ane (aneurine), arg (arginine), asn (asparagine), cit (citrulline), cyn (cystine), gln (glutamine), gly (glycine), leu(leucine), lys (lysine), orn (ornithine), pan (pantothenic acid), pn (pyridoxine), pro (proline), rib (riboflavine), ser (serine), ura (uracil)

2) + (vegetatively compatible), - (vegetatively incompatible)

3) Pro (prototroph), Auxo (auxotroph)

basidiocarps (Kiguchi & Yanagi, 1985).

3.1.2. Genetic recombination. Interspecific fusion products of *Pleurotus* were analysed with respect to the distribution of progenies and segregation of markers by random spore analysis. The genetic characters were shown to segregate and recombine in the first segregation of monosporus isolates taken from basidiocarps of interspecific fusants (Table 4). The analysis provides proof of heterokaryosis, and strong evidence for haploidy of vegetative nuclei, a sexual cycle consisting of nuclear fusion and meiosis. Basidiospores could yield progeny of four genotypes in the crosses ASI 2-29-ade x ASI 2-3-rib, ASI 2-3-rib x ASI 2-1-arg, ASI 2-4-ribo x ASI 2-1-arg and ASI 2-3-ribo x ASI 2-44-lys; i.e. auxotrophs of one parental type, auxotrophs of the other parental type, prototrophs and double auxotrophs, respectively. However, in the three or four factor crosses, segregants were not detected clearly. Comparatively large numbers of prototrophic recombinants were recovered from almost every type of fusant. A 1:1 ratio of allelic loci could be expected from the three crosses, ASI 2-29-ade x ASI 2-3-ribo, ASI 2-3-ribo x ASI 2-1-arg and ASI 2-4-ribo x ASI 2-1-arg. However, this ratio would change to 3:1 with increasing proportions of ade, 2-3-ribo, and 2-4-ribo, respectively. In crosses involving three, four, five or six factors, the ratio of allelic loci was different from the expected ratio based on independent segregation. Most of the parental genotypes were recovered except some fusants. Meiotic segregation and recombination was detected by random basidiospore analysis in thirty-three fusion products from nineteen crosses. The aberration ratio indicated the gene interaction resulting from differences in genome structure between species. In all cases, the germination rate of the basidiospores constituted a very significant feature in the selection of particular genotypes (Raper *et al.*, 1972). Thus, in this situation, it is possible that the prototroph could have a selective advantage (Santiago, 1981).

Clamp connections were first described by Hoffman (1856) and soon thereafter were noted by several authors among the characteristics of basidiomycotina (De Bary, 1866; Hartig, 1866). In heterothallism the mycelium of monosporus isolates lacked clamp connections and was cross-fertile. When monosporus isolates were crossing, dikaryotic hyphae formed clamp connections and were fertile (Raper, 1966; Chang & Miles, 1989). The prototrophic colonies of single spore cultures from the fruit body of somatic hybrids formed clamp connections in the fusion pairings of *P. florida* and *P.*

ostreatus (Yoo *et al.*, 1986). Among the 16 fusants, 3.7-78.5% of the prototrophs tested formed clamps. Those basidiospores that had clamp connections could be mononucleate or binucleate as in cases of homothallism. However, it is not known whether this is the case for protoplast fusion products in all *Pleurotus* species. In strain P5 especially, 78.5% of the prototrophic recombinants tested formed clamp connections. Formation was associated with non-reciprocal recombination. When a diploid cell undergoes meiosis to produce four haploid cells, exactly half of the genes in these cells should be maternal and the other half paternal. This phenomenon was believed to be a straightforward consequence of the mechanisms of general recombination and DNA repair.

3.1.3. Fruit body production. Fruit body production by 40 interspecific hybrids obtained by fusion of protoplasts from *P. florida* ASI 2016 and *P. ostreatus* ASI 2018 was examined on trays of fermented and pasteurized rice straw. The fruit body yield indices of *P. florida-ostreatus* hybrids ranged between 27-155 compared with parental values of 100 (ASI 2018) and 138 (ASI 2016). Eight hybrid strains bore only small numbers of basidiospores or none at all. The fruiting yields of these sporeless strains were lower than parental values. Other breeding programmes were performed to improve varieties with high yields of good quality fruit bodies. The hybrid P72 was derived from 38 protoplast fusion products between *P. florida-ostreatus* recombinant P5-M 43-arg ribo and *P. ostreatus* ASI 2001 (Yoo, 1988). The yield indices of the 38 hybrids ranged between 44-153 compared with the parental values of 100 (ASI 2018), 108 (ASI 2001), and 130 (ASI 2016) respectively. This *P. florida-ostreatus-ostreatus* hybrid P72 was characterized by the formation of a large fruit body of semispherical shape with long stipe and circular pileus. This hybrid was designated as "Wonhyeongneutaribeosus" in Korea, which means a prototype spherical oyster mushroom, and has been distributed to local farmers since 1990. A significant increase in carpophore production was observed in somatic hybrids of protoplasts. This phenomenon may be associated with heterosis due to gene interaction of nucleus and/or mitochondria.

3.2. Intergenus Protoplast Fusion

For the induction of auxotrophic mutants from mycelial fragments of

Elfvingia applanata (syn. *Ganoderma applanatum*) and *Ganoderma lucidum*, ultraviolet light was the mutagenic agent of choice. The following 38 mutants were identified: 18 requiring vitamins, 9 requiring amino acids and 11 requiring nucleic acid bases and related compounds (Park *et al.*, 1987a; Park, 1988; Um *et al.*, 1988). Intergenus fusion products of protoplasts were obtained by polyethylene glycol induced fusion of protoplasts from auxotrophic mutants of *E. applanata* and *G. lucidum*. The fusants produced sectors of a yellowish red-pigmented mycelium on hypertonic MM. Prolonged culture of the fusion products resulted in the pigmentation of mycelial colonies. This melanin pigmentation could be associated with tyrosine metabolism, as reported earlier (Trias *et al.*, 1989). When transferred to MM plates, colonies exhibited regular and irregular shaped appearance. On CM the fusants were classified into stable heterokaryons and spontaneous segregants. After three subcultivations, the colonies morphology of these stable heterokaryons altered to a morphology characteristic of *E. applanata*, *G. lucidum* or an intermediate phenotype. A comparison of intergenus protoplast fusion

TABLE 5. Frequency distribution of progenies of somatic hybrids between *Pleurotus ostreatus* ASI 2-1-arg and *Ganoderma applanatum* ASI 7-18-cyn met following protoplast fusion (Y. B. Yoo unpublished results).

Phenotype[a)]	No. of individuals		
	P382	P386	P399
prototroph	290	204	230
arg	13	1	5
met	0	0	4
arg cyn	0	0	3
arg cyn met	0	1	0
ane	0	1	1
rib	0	1	2
ane rib	0	88	55

a) Mutant symbols; ane (aneurine), arg (arginine), cyn (cystine), met (methionine), rib (riboflavine) products was made using isozyme analysis of esterase. The esterase banding patterns of type 1 fusants could be characterized by new active bands. Type 2 fusion products were similar to those of the *E. applanata* or *G. lucidum* type. Comparison of the fusants showed that interaction had occurred between the two genomes in the fusion progenies. Intergeneric somatic hybrids produced fruit bodies characteristic of *E. applanata, G. lucidum* or intermediate type on oak sawdust-rice bran medium in a glass bottle. The rate of spontaneous segregation in intergenus hybrids was much higher than in interspecies protoplast fusion products. However, no segregants were found in intraspecific protoplast fusion products (Park, 1988). It is suggested that spontaneous segregation of complemented colonies takes place due to the imbalance in the cell contents of the mycelium, particularly in the types and number of nuclei (Yoo *et al.*, 1989c). Even when tested on various media, the germination frequency of basidiospores derived from *E. applanata* and *G. lucidum* was either very poor or zero. For this reason, it was difficult to clarify segregation and genetic recombination in progenies of intergeneric hybrids.

3.3. Interorder Protoplast Fusion

Interorder heterokaryons were obtained by protoplast fusion between *Pleurotus ostreatus* in the order Agaricales and *Ganoderma applanatum* in the order Aphyllophorales. The fusion products of protoplasts were produced after 25-30 days of incubation on MM plates. When transferred to MM plates, all fusion colonies exhibited an extremely slow growth rate. Over three consecutive subcultivations on CM, fusants showed a progressively increasing growth rate. The morphology of 36 fusion products was classified into 5 types: 9 of mixed parental morphology on *Ganoderma* complete medium (GCM; Choi, *et al.*, 1989), 1 of mixed parental morphology on mushroom complete medium (MCM; Raper *et al.*, 1972), 17 of mixed parental morphology on GCM and MCM, 3 stable *P. ostreatus* types, and 6 stable non-parental types. Among the fusant products, seventy five percent showed mixed morphologies or spontaneous segregants of both parents on the first subcultivation on MCM and GCM. The phenotype of these fusants resembled *P. ostreatus* after three subcultivations. The phenotype of the 25% of stable strains did not alter after subcultivation. A comparison of fusants was made using isozyme analysis of esterase, malate dehydrogenase and peroxidase. In most cases, the enzyme patterns of *G. applanatum* were not distinct. However, fusants showed non-parental bands. Hyphae of all fusion products except two strains did not form clamp connections on CM agar or in a liquid CM. Two clamped and three clampless fusants produced mature fruiting bodies on sawdust rice bran medium in a glass bottle. Clamp connections were present in all of these basidiocarps.

Interorder somatic hybrids of *P. ostreatus* and *G. applanatum* were analysed with respect to the distribution of markers and genetic recombination among progenies by random spore analysis (Table 5). The genetic markers were shown to segregate and recombine abnormally in the first segregation of monosporus isolates taken from the fruit body. The genotypes could not be detected in a large number of auxotrophic progenies. The modified Holliday method was used for identification of abnormal progenies. Surprisingly, these strains were aneurine-requiring, riboflavine-requiring, and aneurine and riboflavine-requiring mutants. When parental auxotrophs of *P. ostreatus* ASI 2-1-arg and *G. applanatum* ASI 7-18-cyn met were transferred to MM plus aneurine, MM plus riboflavine, and MM plus aneurine and riboflavine, respectively, they were non-viable. These results indicate that, after fusion, the nuclear and cytoplasmic genomes reassort in a cell. When the two protoplasts first fused, a heterokaryon was produced which contained both nuclei and both cytoplasms. Almost all of these did not remain together and eventually the mononuclear synkaryon cell was formed following fusion of the two nuclei. The genetic information from *P. ostreatus* chromosomes was predominant in the synkaryon . However, most of the fusants retained genes from both parents. Fusion products between *Pleurotus sajor-caju* and *Schizophyllum commune* have also been described by Liang & Chang (1989). Most fusants were abortive and died early, while some strains produced very small ugly callus tissue or normal fruit bodies similar to *P. sajor-caju*. These authors suggested that the fusants were not real heterokaryons due to asynchronous physiological development of the two partners.

4. TRANSFER OF SPORE AND CELL ORGANELLES INTO PROTOPLASTS

4.1. Oidium Transfer

TABLE 6. Characteristics of somatic hybrids by nuclear transfer in higher fungi.

Strain[1)] (D + R)	Clamp[2)] Anasto-mosis	Clamp[2)] Nuclear transfer	Fruiting[3)] Type	Fruiting[3)] Clamp	Genetic background	Reference[4)]
P. f-m + P. o-m	+	+	intermediate	+	heterokaryon	Yoo *et al.* (1987e)
		-	non-fertile		nuclear hybrid	
		-	non-fertile		homokaryotic recombinant	
P. o-m + P. f-m	+	+	intermediate	+	heterokaryon	
P. sd-d + P. o-m	-	+	*P. ostreatus*	+	heterokaryon	after You *et al.* (1988a)
		-	non-fertile		heterokaryon	
A. a-d + P. f-m	-	+	*A. aegerita*	+	heterokaryon	Yoo *et al.* (1989b)
		-	*A. aegerita*	-	heterokaryon	
		-	*A. aegerita*	-	reconstituted cell	
L. e-d + P. f-m	-	+	*L. edodes*	+	reconstituted cell	
		-	*P. florida*	+	synkaryon	
		-	non-fertile		synkaryon	
L. e-d + P.ss-d	-	+	*P. salmoneo-stramineus*	+	heterokaryon	
E. a-d + P. o-m	-	-	*P. ostreatus*	+	synkaryon	
		-	non-fertile	+	synkaryon	

1) D (donor), R (recipient), A. a (*Agrocybe aegerita*), E. a (*Elfvingia applanata*), L. e (*Lentinus edodes*), P. f (*Pleurotus florida*), P. o (*P. ostreatus*), P. sd (*P. sapidus*), P. ss (*P. salmoneostramineus*), m (monokaryon), d (dikaryon)
2), 3) + (Present clamp connection), - (Absent clamp connection)
4) Y. B. Yoo unpublished results.

In any discussion of the conversion of spores into protoplasts, the question arises as to what extent the two different species can fuse in such affinity. After polyethylene glycol solution treatment of the mixture of donor oidia of

TABLE 7. Segregation and recombination of genetic characters in progenies of somatic hybrids by nuclear transfer (Y. B. Yoo unpublished results).

Strain[a)] Donor(A)	Recipient(B)	Fusant	Compatibility[b)]	Parentals A	Parentals B	Recombinant[c)] Pro.	Recombinant[c)] Auxo.	% of recomb.
P. florida	+ *P. ostreatus*	P421	+	44	34	10	2	13.33
2-3-rib	2-1-arg	P425		49	0	64	7	59.16
P. ostreatus	+ *P. florida*	P302	+	0	56	54	9	52.94
2-1-arg	2-3-rib	P312		48	32	14	9	22.33
P. sapidus	+ *P. ostreatus*	P616	-	269	51			
2057-w	2-1-arg	P617		248	72			
L. edodes	+ *P. florida*	P664	-	69	51			
3046-w	2-3-rib	P673		43	133			

a) Mutant symbol : arg (arginine), rib (riboflavine), w (wild)
b) + (vegetatively compatible), - (vegetatively incompatible)
c) Pro (prototroph), Auxo (auxotroph)

Lyophyllum ulmarium ASI 8007-wild (syn. *Hypsizigus marmoreus*) in the Agaricales and recipient protoplasts of *Ganoderma applanatum* ASI 7-18-cyn met in the Aphyllophorales, small colonies appeared on the minimal medium plates (Yoo, 1989). In order to check for back mutation, the protoplasts from *G. applanatum* control plates were inoculated onto MM. After 5 days-culturing, protoplasts reverted to mycelial colonies. When transferred to GCM plates, fusants showed mixed morphologies of both of the parents. During three subsequent subcultivations the morphology of fusants changed into a type similar to that of *L. ulmarium*. All fusion products produced oidia, clamp connections and fruiting bodies similar to those of *L. ulmarium*. Esterase isozyme pattern of interorder fusants showed both pre-existing and new bands. Each individual fusion product showed no difference in mycelial growth rate, colony morphology, esterase band pattern and basidiocarp.

In the chance of fusion between hypha and oidium, some researchers reported on intraspecies and interspecies (Bistis, 1970; Fries, 1981; Ingold,

1984). In this case, however, interorder fusion between reverted hyphae of *G. applanatum* and oidia of *L. ulmarium* was not possible because of a vegetative incompatibility. These results signified that fusion products between protoplasts and oidia were extremely stable compared to fusants formed by other gene transfer method such as protoplast fusion, cell organelle transfer and transformation.

4.2. Nucleus Transfer

Membrane-bound nuclei (karyoplasts or miniprotoplasts) has been isolated using chemicals (Lörz & Potrykus, 1978; Becher *et al.*, 1982; Ferenczy & Pesti, 1982; Willmitzer, 1984; Saxena *et al.*, 1986) and by centrifugation (Lörz & Potrykus, 1980; Lörz *et al.*, 1981). The transfer of isolated nuclei from protoplasts and mycelia into protoplasts has been studied in higher fungi (Table 6). The compatible transfer of isolated nuclei from *P. florida* into protoplasts of *P. ostreatus* was induced with polyethylene glycol and $CaCl_2$. The transfer products of nuclei were classified into three types: nuclear hybrid or allodiploid, heterokaryon and homokaryotic recombinant. The nuclear hybrid produced more vigorous mycelial growth and was stable on CM. One of the hybrid colonies appeared segregated on CM plus benomyl. The hyphae neither formed clamp connections nor developed fruiting bodies. The heterokaryon type was the main product of nuclear transfer. The colony formed clamp connections and was fertile on sawdust substrates. The last type was very slow growing or were non-viable after fragments of nuclei or chromosomes were transferred into recipient protoplasts. Clamps and primordia were not produced due to the homokaryotic character of the strain. Isozyme patterns of esterase in the allodiploid revealed a new band. Heterokaryotic and homokaryotic products could be characterized by parental bands. The genetic markers were shown to segregate and recombine in the first generation of monosporus isolates from the fruiting bodies of the nuclear transfer products (Table 7). The recombination rate of these products was low compared with that of protoplast fusants. In the incompatible transfer of nuclei from *P. sapidus* (wild) mycelia into protoplasts of *P. ostreatus* (arg), spontaneous segregants from both parents, non-parental segregants and stable heterokaryons were obtained.

The intergeneric nuclear transfer products arising from donor nuclei of *Lentinus edodes* (wild) and recipient protoplasts of *P. florida* (ribo) were nuclear hybrids, synkaryons, and reconstituted cells. Nuclear hybrids grew more vigorously and were stable on CM. This heteroploid or allodiploid was non-fertile. Synkaryons did not form clamps on CM agar or in a liquid CM. Primordia were induced by a light-dark cycle during growth on sawdust rice bran medium in a glass bottle. They developed the clamped fruiting bodies similar to those of *P. florida*. The reconstituted cells with clamp connections developed into the basidiocarps similar to those of *L. edodes*. In this combination there were two genotypes one of which was the wild-type and the other a riboflavine-requiring auxotroph.

Interfamily products were obtained following transfer of nuclei from *A. aegerita* (wild) into protoplasts of *P. florida* (ribo$^-$). Type 1 products consisted of 3 spontaneous segregants exhibiting both parental forms of colony morphology. Clamp connections were present in hyphae of the *A. aegerita* kind but were lacking in *P. florida*-type hyphae. Type 2 consisted of 55 clamped heterokaryons, and type 3 of 42 clampless products. The monokaryon of type 3 products was composed of reconstituted cells derived from the donor karyoplast and the enucleate recipient cytoplast. All interfamily products produced primordia and developed fruit bodies similar to those of *A. aegerita* on CM agar or sawdust medium.

Interspecific and intergeneric nuclear hybrids were derived from prolonged cultivation of the heterokaryon on MM. These nuclear hybrid colonies were distinguished from heterokaryons because they produced more compact and more vigorously growing mycelium. When cultured on CM plus haploidizing agents, the allodiploid appeared to be in segregation. The growth characteristics of the hybrids were similar to those observed for *Aspergillus* (Kevei & Peberdy, 1977, 1979), *Penicillium* (Anne *et al.*, 1976; Peberdy, *et al.*, 1977; Anne, 1982) and yeast hybrids (Ferenczy, 1985). Segregants were obtained from hybrids grown in the presence of benomyl, chloral hydrate and paraflurophenylalanine (Kevei & Peberdy, 1979; Mellon *et al.*, 1983). These hybrids were also identified by various criteria including the secretion of brown pigment, doubled DNA content, formation of few conidia, and large spore size. The white mycelium of the hybrids did not form clamp connections and was non-fertile. As reported by other workers (Anne *et al.*, 1976; Peberdy *et al.*, 1977; Anne & Eyssen, 1978), no pigmentation occurred on MM and CM.

4.3. Chromosome Transfer

Studies of chromosome-mediated gene transfer have been reported in animal cells (McBride & Dzer, 1973; Burch & McBride, 1975; Klobutcher *et al.*, 1980; Klobutcher & Ruddle, 1981) and higher plants (Szabados *et al.*, 1981; Griesbach *et al.*, 1982). The uptake of isolated chromosomes from *Lyophyllum ulmarium* by protoplasts of *Ganoderma applanatum* was induced with polyethylene glycol. Products of chromosome uptake by protoplasts showed microtransgenome and macrotransgenome types (Yoo *et al.*, 1988b). The former was slow-growing and exhibited irregular sectoring on CM, and the latter produced an outgrowing and stable mycelial colony composed of thick hyphae and was segregated on CM plus benomyl. A comparison of macrotransgenome types was attempted using isozyme analysis of esterase. The enzyme pattern of the transformants was very distinct from that of their parents in both position and quantity. Transformants produced after uptake of chromosomes from *Pleurotus florida* by protoplasts of *P. ostreatus* showed only microtransgenome type (Yoo, 1988). All types of transformants did not form clamps and were non-fertile.

5. GENETIC INFORMATION TRANSFER BY HEAT-INACTIVATED PROTOPLASTS

The employment of wild types as fusion partners could greatly enhance mushroom breeding research programs. Inactivated protoplasts as fusion partners have been studied in *Bacillus* (Levi *et al.*, 1977; Fodor *et al.*, 1978), *Micromonospora* (Szvoboda *et al.*, 1980), *Streptomyces* (Hopwood & Wright, 1981; Ochi, 1982; Baltz & Matsushima, 1983) and *Aspergillus* (Ferenczy, 1984). Protoplasts of *Pleurotus florida* ASI 2-3-ribo, incubated at 60C° for 60 min, lost the ability to revert to a mycelial colony. Such heat-inactivated protoplasts, however, gave rise to recombinants when they were fused by polyethylene glycol treatment with normal protoplasts from *P. ostreatus* ASI 2-1-arg (Y. B. Yoo, unpublished results). These transformants were slow growing and developed irregular sectoring colonies similar to those of microtransgenome types resulting from chromosome uptake. The mycelial colonies did not form clamps and were non-fertile.

6. TRANSFORMATION

There are two types of vector for transformation of fungi. The naked DNA donor can establish itself as an autonomously replicating sequence or integrate into the host chromosome after entry into the recipient cell. Investigations of DNA-mediated transformation have been reported in higher fungi (Banks, 1983; Munoz-Rivas *et al.*, 1986; Binninger *et al.*, 1987; Mellon *et al.*, 1987; Specht *et al.*, 1988; Alic *et al.*, 1989; Burrows *et al.*, 1990). *Pleurotus florida* Leu 2 auxotroph, deficient in β-isopropylmalate dehydrogenase, was transformed to Leu+ with plasmid pM 301 DNA containing the *Flammulina velutipes* sequence (Byun *et al.*, 1989a). A uracil auxotroph of *Pleurotus florida* P101 was also transformed to prototrophy by using a chimeric vector containing *Aspergillus nidulans* ans 1, and *Neurospora crassa* pyr 4 DNA (Byun *et al.*, 1989b). Southern hybridization revealed that the transforming DNA was integrated into the chromosomal DNA. Transformants grew very slowly on MM and the hyphae did not form clamp connections. Recently, Mutasa *et al.* (1990) reported the formation of unfused clamp cells in a monokaryotic strain following transformation into a compatible host strain of *Coprinus cinereus*.

7. CONCLUDING REMARKS

Nonconventional methods of genetic manipulation, such as protoplast fusion, cell organelle transfer and transformation, have been carried out on several higher fungi. However, studies on genome structure have not been conducted widely in economically important mushrooms. Protoplast technology should play a dominant role in the development of genetics and breeding in mushrooms. It is important to note that strain improvement by protoplast fusion has been performed in closely related species. Successful recombination between distantly related species provides the opportunity for a broadening of the gene pool. The uptake of isolated cell organelles by protoplasts may allow the exploitation of the protoplast fusion technique on a more refined level. DNA-mediated transformation related to gene vectors advances our understanding of the basic molecular biology of higher fungi. Development of this method also opens up many possibilities for the commercial production of medically important substances found in higher fungi.

REFERENCES

ALIC, M., KORNEGAY, J. R., PRIBNOW, D. & GOLD, M. H. (1989). Trans- formation by complementation of an adenine auxotroph of the lignin-degrading basidiomycete *Phanerochaete chrysosporium. Applied and Environmental Microbiology* **55**, 406-411.

ANDERSON, N. A., WANG, S. S. & SCHWANDT, J. W. (1973). The *Pleurotus ostreatus-sapidus* species complex. *Mycologia* **65**, 28-35.

ANNE, J. (1982). Comparison of penicillins produced by interspecies hybrids from *Penicillium chrysogenum. European Journal of Applied Microbiology and Biotechnology* **15**, 41-46.

ANNE, J. & EYSSEN, H. (1978). Isolation of interspecies hybrids of *Penicillium citrinum* and *Penicillium cyaneofulvum* following protoplast fusion. *FEMS Microbiology Letters* **4**, 87-90.

ANNE, J., ESSEN, H. & DE SOMER, P. (1976). Somatic hybridization of *Penicillium roquefortii* with *P. chrysogenum* after protoplast fusion. *Nature* **262**, 719-721.

ASCHAN, K. (1952). Dediploidisation mycelia of the basidiomycete *Collybia velutipes. Svensk Botanisk Tidskrift* **46**, 366-392.

BALTZ, R. H. & MATSUSHIMA, P. (1983). Advances in protoplast fusion and transformation in *Streptomyces.* In *Protoplasts 1983. Lecture Proceedings.* pp. 143-148. Edited by I. Potrykus, C. T. Harms, A. Hinnen, R. Hutter, P. J. King, & R. D. Shillito. Basel: Birkhäuser Verlag.

BANKS, G. R. (1983). Transformation of *Ustilago maydis* by a plasmid containing yeast 2-micron DNA. *Current Genetics* **7**, 73-77.

BECHER, D., CONRAD, B. & BOTTCHER, F. (1982). Genetic transfer mediated by isolated nuclei in *Saccharomyces serevisiae. Current Genetics* **6**, 163-165.

BINNINGER, D. M., SKRZYNIA, M. C., PUKKILA, P. J. & CASSELTON, L. A. (1987). DNA-mediated transformation of the basidiomycete *Coprinus cinereus. EMBO Journal* **6**, 835-840.

BISTIS, G. N. (1970). Dikaryotization in *Clitocybe truncicola. Mycologia* **62**, 911-924.

BOK, J. W., PARK, S. H., CHOI, E. C., KIM, B. K. & YOO, Y. B. (1990). Studies on protoplast formation and regeneration of *Coriolus versicolor. Korean Journal of Mycology* **18**, 115-126.

BURCH, J. W. & MCBRIDE, O. W. (1975). Human gene expression in rodent cells after uptake of isolated metaphase chromosomes. *Proceeding of the National Academy of Sciences of the United States of America* **72**, 1797-1801.

BURROWS, D. M., ELLIOTT, T. J. & CASSELTON, L. A. (1990). DNA-mediated transformation of the secondarily homothallic basidiomycete *Coprinus bilanatus*. *Current Genetics* **17**, 175-177.

BYUN, M. O., YOO, Y. B., GO, S. J., YOU, C. H., CHA, D. Y. & PARK, Y. H. (1989a). Transformation of the β-isopropylmalate dehydrogenase gene of *Flammulina velutipes* into *Pleurotus florida*. *Korean Journal of Mycology* **17**, 27-30.

BYUN, M. O., YOO, Y. B., YOU, C. H., CHA, D. Y. & CHO, M. J. (1989b). Transformation of *Pleurotus florida* with *Neurospora* pyr 4 gene. *Korean Journal of Mycology* **17**, 209-213.

CHANG, S. T. & MILES, P. G. (1989). *Edible Mushrooms and Their Cultivation*. Boca Raton: CRC Press.

CHOI, S. H., KIM, B. K., KWAK, J. H., CHOI, E. C., KIM, Y. C., YOO, Y. B. & PARK, Y. H. (1987). Studies on protoplast formation and regeneration of *Ganoderma lucidum*. *Archives of Pharmaceutical Research* **10**, 158-164.

DAVIS, B. (1985). Factors influencing protoplast isolation. In *Fungal Protoplasts : Application in Biochemistry and Genetics*, pp. 45-71. Edited by J. F. Peberdy & L. Ferenczy. New York: Marcel Dekker.

DE BARY, A. (1886). *Morphologia und Physiologie der pilze, Flechten, und Myxomyceten*. Leipzig: USW.

DE VRIES, M. H. & WESSELS, J. G. H. (1972). Release of protoplasts from *Schizophyllum commune* by a lytic enzyme preparation from *Trichoderma viride*. *Journal of General Microbiology* **73**, 13-22.

EGER, G. (1974). Rapid method for breeding *Pleurotus ostreatus*. *Mushroom Science* **9**, 567-573.

ELLIOTT, T. J. (1985). Developmental genetics from spore to sporophore. In *Developmental Biology of Higher Fungi*, pp. 451-465. Edited by D. Moore, L. A. Casselton, D. A. Wood & J. C. Frankland. Cambridge: Cambridge University Press.

EUGENIO, C. P. & ANDERSON, N. A. (1968). The genetics and cultivation of *Pleurotus ostreatus*. *Mycologia* **60**, 627-634.

FARKAS, V. (1985). The fungal cell wall. In *Fungal Protoplasts: Application in Biochemistry and Genetics*, pp. 3-29. Edited by J. F.

Peberdy & L. Ferenczy. New York: Marcel Dekker.

FERENCZY, L. (1984). Fungal protoplast fusion: basic and applied aspects. In *Cell Fusion : Gene Transfer and Transformation*, pp. 145- 169. Edited by R. F. Beers Jr. & E. G. Bassett. New York: Raven Press.

FERENCZY, L. (1985). Protoplast fusion in yeasts. In *Fungal Protoplasts. Applications in Biochemistry and Genetics*, pp. 279-386. Edited by J. F. Peberdy & L. Ferenczy. New York: Marcel Dekker.

FERENCZY, L. & PESTI, M. (1982). Transfer of isolated nuclei into protoplasts of *Saccharomyces cerevisiae. Current Microbiology* **7**, 157-160.

FERENCZY, L., KEVEI, F. & SZEGEDI, M. (1975). Increased fusion frequency of *Aspergillus nidulans* protoplasts. *Experientia* **31**, 50-52.

FINCHAM, J. R. S. (1989). Transformation in fungi. *Microbiological Reviews* **53**, 148-170.

FODOR, K., DEMIRI, E. & ALFOLDI, L. (1978). Polyethylene glycol-induced fusion of heat-inactivated and living protoplasts of *Bacillus megaterium. Journal of Bacteriology* **135**, 68-70.

FRIES, N. (1981). Recognition reactions between basidiospores and hyphae in *Lecinum. Transaction of the British Mycological Society* **77**, 9-14.

GINTEROVA, A. (1973). Dedikaryotization of higher fungi in submerged culture. *Folia Microbiologia* **18**, 277-280.

GO, S. J. & SHIN, G. C. (1986). Incompatibility factors and genetic analysis of *Pleurotus sajor-caju. Korean Journal of Mycology* **14**, 17-23.

GO, S. J., CHA, D. Y. & PARK, Y. H. (1981). Intra- and intermatings among strains of *Pleurotus ostreatus* and *P. florida. Korean Journal of Mycology* **9**, 13-18.

GO, S. J., SHIN, G. C. & YOO, Y. B. (1985). Protoplast formation, regeneration and reversion in *Pleurotus ostreatus* and *P. sajor-caju. Korean Journal of Mycology* **13**, 169-177.

GO, S. J., YOU, C. H. & SHIN, G. C. (1989). Effects of incompatibility on protoplast fusion between intra- and interspecies in basidiomycete, *Pleurotus* spp. *Korean Journal of Mycology* **17**, 137-144.

GRIESBACH, R. J., MALMBERG, R. L. & CARLSON, P. S. (1982). Uptake of isolated lily chromosomes by tobacco protoplast. *Journal of Heredity* **73**, 151-152.

HAMLYN, P. F., BRADSHAW, R. E., MELLON, F. M., SANTIAGO, C. M., WILSON, J. M. & PEBERDY, J. F., Efficient protoplast isolation from fungi using commercial enzymes. *Enzyme and Microbial Technology* **3**, 321-325.

HARDER, R. (1927). Zur Frage der Rolle von Kern und Protoplasma in Zellgeschehen und bei der Übertragung von Eigenschaften. *Zeitschrift für Botanik.* **19**, 337-407.

HARTIG, T. (1866). Wichtige Krankheiten der Waldbaume Berlin: Springer.

HILBER, O. (1982). Die Gattung Pleurotus (Fr.) Kummer unter besonderer Berñcksichtigung des Pleurotus eryngii Formenkomplexes. *Bibliotheca Mycologica* **87**. Vadus: Cramer.

HOFFMAN, H. (1856). Die Pollinarien und Spermatien von *Agaricus. Botanische Zeitung* **14**, 137-148.

HOPWOOD, D. A. & WRIGHT, H. M. (1981). Protoplast fusion in *Streptomyces*: Fusions involving ultraviolet-irradiated protoplasts. *Journal of General Microbiology* **126**, 21-27.

INGOLD, C. T. (1984). Fusion of hyphae of *Pleurotus ostreatus* with its ungerminated basidiospores. *Transaction of the British Mycological Society* **83**, 724-725.

KERRUISH, R. M. & DACOSTA, E. W. B. (1963). Monocaryotization of cultures of *Lenzites trabea* (Pers.) Fr. and other wood-destroying basidiomycetes by chemical agents. *Annals of Botany* **27**, 653-670.

KEVEI, F. & PEBERDY, J. F. (1977). Interspecific hybridization between *Aspergillus nidulans* and *Aspergillus rugulosus* by fusion of somatic protoplasts. *Journal of General Microbiology* **102**, 255-262.

KEVEI, F. & PEBERDY, J. F. (1979). Induced segregation in interspecific hybrids of *Aspergillus nidulans* and *Aspergillus rugulosus* obtained by protoplast fusion. *Molecular and General Genetics* **170**, 213-218.

KIGUCHI, T. & YANAGI, S. O. (1985). Intraspecific heterokaryon and fruit body formation in *Coprinus macrorhizus* by protoplast fusion of auxotrophic mutants. *Applied Microbiology and Biotechnology* **22**, 121-127.

KLOBUTCHER, L. A., MILLER, C. L. & RUDDLE, F. H. (1980). Chromosome- mediated gene transfer results in two classes of unstable transformants. *Proceeding of the National Academy of Sciences of the United Stares of America* **77**, 3016-3614.

KLOBUTCHER, L. A. & RUDDLE, F. H. (1981). Chromosome-mediated gene transfer. *Annual Review of Biochemistry* **50**, 533-534.

LEE, Y. H., PARK, Y. H., Y00, Y. B. & MIN, K. H. (1986a). Studies on protoplast isolation of *Pleurotus cornucopiae*. *Korean Journal of Mycology* **14**, 141-148.

LEE, Y. H., YOU, C. H., CHA, D. Y., YOO, Y. B. & MIN, K. H. (1986b). Protoplast regeneration and reversion in *Pleurotus cornucopiae*. *Korean Journal of Mycology* **14**, 215-223.

LEVI, C., SANCHEZ RIVAS, C. & SCHAEFFER, P. (1977). Further genetic studies on the fusion of bacterial protoplasts. *FEMS Microbiology Letters* **2**, 323-326.

LIANG, Z. R. & CHANG, S. T. (1989). A study on intergeneric hybridization between *Pleurotus sajor-caju* and *Schizophyllum commune* by protoplast fusion. *Mushroom Science* **12**, 125-137.

LÖRZ, H. & POTRYKUS, I. (1978). Investigations on the transfer of isolated nuclei into plant protoplasts. *Theoretical and Applied Genetics* **53**, 251-256.

LÖRZ, H. & POTRYKUS, I. (1980). Isolation of subprotoplasts for genetic manipulation studies. In *Advances in Protoplast Research*, pp. 377-382. Edited by L. Ferenczy & G. L. Farkas. Oxford: Pergamon Press.

LÖRZ, H., PASZKOWSKI, J., DIEKERKS-VENTLING, C. & POTRYKUS, I. (1981). Isolation and characterization of cytoplasts and miniprotoplasts derived from protoplasts of cultured cells. *Physiologia Plantarum* **53**, 385-391.

MAY, B., HENLEY, K. J., FISHER, C. G. & ROYSE, D. J. (1988). Linkage relationships of 19 allozyme encoding loci within the commercial mushroom genus *Pleurotus*. *Genome* **30**, 888-895.

MCBRIDE, D. W. & DZER, H. L. (1973). Transfer of genetic information by purified metaphase chromosomes. *Proceeding of the National Academy of Sciences of the United States of America* **70**, 1258-1262.

MELLON, F. M., LITTLE, P. R. R. & CASSELTON, L. A. (1987). Gene cloning and transformation in the basidiomycete fungus *Coprinus cinereus*: isolation and expression of the isocitrate lyase gene (acu-7). *Molecular and General Genetics* **210**, 352-357.

MELLON, F. M., PEBERDY, J. F. & MACDONALD, K. D. (1983). Hybridization of *Penicillium chrysogenum* and *Penicillium baarnense* by protoplast fusion: genetic and biochemical analysis. In *Protoplasts* 1983, *Poster Proceedings*, pp. 310-311. Edited by I. Potrykus, C. T. Harms, A. Hinnen, R. Hütter, P. J. King & R. D. Shillito. Basel: Birkhäuser Verlag.

MILES, P. G. & RAPER, J. R. (1956). Recovery of the component strains from dikaryotic mycelia. *Mycologia* **48**, 484-494.

MUNOZ-RIVAS, A., SPECHT, C. A., DRUMMOND, B. J., FROELIGER, E., NOBOTNY, C. P. & ROBERT, C. U. (1986). Transformation of the basidiomycete, *Schizophyllum commune*. *Molecular and General Genetics* **205**, 103-106.

MUTASA, E. S., TYMON, A. M., GÖTTGENS, B., MELLON, F. M., LITTLE, P. F. R. & CASSELTON, L. A. (1990). Molecular organisation of an A mating type factor of the basidiomycete fungus *Coprinus cinereus*. *Current Genetics* **18**, 223-229.

OCHI, K. (1982). Protoplast fusion permits high-frequency transfer of a *Streptomyces* determinant which mediates actinomycin synthesis. *Journal of Bacteriology* **150**, 592-597.

PARAG, A. (1961). Sensitivity of *Schizophyllum commune* to chemical toxicants. *Canadian Journal of Microbiology* **7**, 838-841.

PARK, S. H. (1988). Studies on protoplast and nuclear fusion of *Ganoderma* species. M.Sc. Thesis, Seoul National University (Korea).

PARK, Y. D., YOO, Y. B., CHA, D. Y., CHANG, M. W. & LEE, J. S. (1987a). Isolation of auxotrophic mutants in *Ganoderma applanatum*. *Korean Journal of Applied Microbiology and Bioengineering* **15**, 230-233.

PARK, Y. D., YOO, Y. B., PARK, Y. H., CHANG, M. W. & LEE, J. S. (1987b). Protoplast formation and reversion of *Ganoderma applanatum*. *Korean Journal of Applied Microbiology and Bioengineering* **15**, 234-240.

PARK, Y. D., YOO, Y. B., SHIN, P. G., YOU, C. H., CHA, D. Y., PARK, Y. H. & LEE, J. S. (1988). Interspecific protoplast fusion of *Ganoderma applanatum* and *Ganoderma lucidum* and fruit body formation of fusants. *Korean Journal of Mycology* **16**, 79-86.

PEBERDY, J. F. (1985). Mycolytic enzymes. In *Fungal Protoplasts: Application in Biochemistry and Genetics*, pp. 31-44. Edited by J. F. Peberdy & L. Ferenczy. New York: Marcel Dekker.

PEBERDY, J. F. (1989). Fungi without coats-protoplasts as tools for mycological research. *Mycological Research* **93**, 1-20.

PEBERDY, J. F. (1990). Fungal cell walls. In *Biochemistry of Cell Walls and Membranes in Fungi*, pp. 5-30. Edited by P. J. Kuhn, A. P. J. Trinci, M. J. Jung, M. W. Goosey & L. G. Copping. Berlin: Springer-Verlag.

PEBERDY, J. F., BUCKLEY, C. E., DALTREY, D. C. & MOORE, P. M.

(1976). Factors affecting protoplast release in some filamentous fungi. *Transactions of the British Mycological Society* **67**, 23-26.

PEBERDY, J. F., EYSSEN, H. & ANNE, J. (1977). Interspecific hybridization between *Penicillium chrysogenum* and *Penicillium cyaneofulvum* following protoplast fusion. *Molecular and General Genetics* **157**, 281-284.

RAPER, J. R. (1966). *Genetics of Sexuality in Higher Fungi*. New York: The Ronald Press Company.

RAPER, C. A., RAPER, J. R. & MILLER, R. E. (1972). Genetic analysis of the life cycle of *Agaricus bisporus*. *Mycologia* **64**, 1088-1117.

ROXON, J. E. & JONG, S. C. (1977). Sexuality of an edible mushroom, *Pleurotus sajor-caju*. *Mycologia* **69**, 203-205.

SAMSUDIN, S. & GRAHAM, K. M. (1984). Monokaryotic fruiting in *Pleurotus flabellatus*. *Malaysian Journal of Applied Biology* **13**, 61-65.

SANTIAGO, C. M. (1981). Studies on the physiology and genetics of *Volvariella volvacea* (Bull. ex. Fr.). SINGER. Ph.D. Thesis, University of Nottingham.

SAXENA, P. K., MII, M., CROSBY, W. L., FOWKE, L. C. & KING, J. (1986). Transplantation of isolated nuclei into plant protoplasts. *Planta* **168**, 29-35.

SHIN, G. C., YEO, U. H., YOO, Y. B. & PARK, Y. H. (1986). Some factors affecting the protoplast formation and regeneration from the mycelium of *Ganoderma lucidum* (Fr.) Karsten. *Research Report of Agricultural Science Technology of Chungnam National University* (Korea) **13**, 185-192.

SPECHT, C. A., MUNOZ-RIVAS, A., NOVOTNY, C. P. & ULLRICH, R. C. (1988). Transformation of *Schizophyllum commune*: An analysis of parameters for improving transformation frequencies. *Experimental Mycology* **12**, 357-366.

SZABADOS, A., HADLACZKY, GY. & DUDITS, D. (1981). Uptake of isolated plant chromosomes by plant protoplasts. *Planta* **151**, 141-145.

SZVOBODA, G., LANG, T., GADO, I., AMBRUS, G., KARI, C., FODOR, K. & ALFÖLDI, L. (1980). Fusion of *Micromonospora* protoplasts. In *Advances in Protoplast Research*, pp. 235-240. Edited by L. Ferenczy & G. L. Farkas. Oxford: Pergamon Press.

TAKEMARU, T. & KAMADA, T. (1971). Gene control of basidiocarp development in *Coprinus macrorhizus*. *Report of Tottori Mycological Institute* (Japan) **9**, 21-35.

TERAKAWA, H. (1960). The incompatibility factors in *Pleurotus ostreatus*. *Scientific Papers of the College of General Education, University of Tokyo* **10**, 65-71.

TOYOMASU, T., MATSUMOTO T. & MORI, K. (1986). Interspecific protoplast fusion between *Pleurotus ostreatus* and *Pleurotus salmoneostramineus*. *Agricultural and Biological Chemistry* **50**, 223-225.

TOYOMASU, T. & MORI, K-I. (1989). Characteristics of the fusion products obtained by intra- and interspecific protoplast fusion between *Pleurotus* species. *Mushroom Science* **12**, 151-159.

TRIAS, T., VINAS, M., GUINEA, J. & LOREN, J. G. (1989). Brown pigmentation in *Serratia marcescens* cultures associated with tyrosine metabolism. *Canadian Journal of Microbiology* **35**, 1037-1042.

UM, S. D., CHAE, Y. A., PARK, Y. H. & YOO, Y. B. (1988). Studies on auxotroph induction of *Ganoderma lucidum* and interspecific protoplast fusion between *G. lucidum* and *G. applanatum*. *Korean Journal of Mycology* **16**, 16-20.

VANDENDRIS, R. (1932). La tetrapolarité sexuelle de *Pleurotus columbinus*. *La Cellule* **41**, 267-279.

VANDENDRIS, R. (1933). De la valeur du barrage sexuelcomme critérium dans l'analyse dúne sporée tetrapolaraire de Basidiomycéte: *Pleurotus ostreatus*. *Genetica* **15**, 202-212.

WESSELS, J. G. H., HOEKSEMA, H. L. & STEMERDING, D. (1976). Reversion of protoplasts from dikaryotic mycelium of *Schizophyllum commune*. Protoplasma **89**, 317-321.

WILLMITZER, L. (1984). Isolation of organelles: nuclei. In *Cell Culture and Somatic Cell Genetics of Plant*, pp. 454-460. Edited by I. K. Vasil. Orlando: Academic Press.

WU, L. C. (1987). Strategies for conservation of genetic resources. In *Cultivating Edible Fungi*, pp. 183-211. Edited by P. J. Wuest, D. J. Royse & R. B. Beelman. Amsterdam: Elsevier.

YEA, U. H., YOO, Y. B., PARK, Y. H. & SHIN, G. C. (1988). Isolation of protoplasts from *Flammulina velutipes*. *Korean Journal of Mycology* **16**, 70-78.

YOO, Y. B. (1988). Protoplast fusion and chromosome transfer in *Pleurotus*. Ph.D. Thesis, Gyeongsang National University (Korea).

YOO, Y. B. (1989). Fusion between protoplasts of *Ganoderma applanatum* and oidia of *Lyophyllum ulmarium. Korean Journal of Mycology* **17**, 197-201.

YOO, Y. B. (1991). Interspecific hybridization between *Pleurotus cornucopiae* and *Pleurotus florida* following protoplast fusion, *Korean Journal of Mycology*. (in press).

YOO, Y. B., BYUN, M. O., GO, S. J., YOU, C. H., PARK, Y. H. & PEBERDY, J. F. (1984). Characteristics of fusion products between *Pleurotus ostreatus* and *Pleurotus florida* following interspecific protoplast fusion. *Korean Journal of Mycology* **12**, 164-169.

YOO, Y. B., PEBERDY, J. F. & YOU, C. H. (1985). Studies on protoplast isolation from edible fungi. *Korean Journal of Mycology* 13(1), 1-10.

YOO, Y. B., YOU, C. H., PARK, Y. H. & PEBERDY, J. F. (1986). Genetic analysis of the life cycle in interspecific hybrids of *Pleurotus ostreatus* and *Pleurotus florida* following protoplast fusion. *Korean Journal of Mycology* **14**, 9-15.

YOO, Y. B., BYUN, M. O., GO, S. J. YOU, C. H. & CHANG, K. Y. (1987a). Protoplast isolation and reversion from *Pleurotus spodoleucus. Korean Journal of Mycology* **15**, 19-22.

YOO, Y. B., LEE, Y. H., YEO, U. H., UM, S. D., CHA, D. Y. & PARK, Y. H. (1987b). Selection of neohaplont in some edible fungi by protoplast reversion. *Korean Journal of Mycology* **15**, 38-41.

YOO, Y. B., YOU, C. H., PARK, Y. H. & CHANG, K. Y. (1987c). Protoplast isolation and reversion from *Lyophyllum ulmarium. Korean Journal of Mycology* **15**, 14-18.

YOO, Y. B., YOU, C. H., PARK, Y. H., LEE, Y. H., CHANG, K. Y. & PEBERDY, J. F. (1987d). Interspecific protoplast fusion and sexuality in *Pleurotus. Korean Journal of Mycology* **15**, 135-141.

YOO, Y. B., YOU, C. H., SHIN, P. G., PARK, Y. H. & CHANG, K. Y. (1987e). Transfer of isolated nuclei from *Pleurotus florida* into protoplasts of *Pleurotus ostreatus. Korean Journal of Mycology* **15**, 250-253.

YOO, Y. B., PARK, Y. H. & CHANG, K. Y. (1988a). Induction of auxotrophic mutants and back mutation in *Pleurotus. Research Report of Rural Development Administration* (*Korea*) **30**, 133-140.

YOO, Y. B., YOU, C. H. & CHANG, K. Y. (1988b). Uptake of isolated *Lyophyllum ulmarium* chromosomes by *Ganoderma applanatum* protoplasts. *Korean Journal of Mycology* **16**, 247-252.

YOO, Y. B., BYUN, M. O., GO, S. J., YOU, C. H. & CHA, D. Y. (1989a).

Protoplast isolation and reversion from *Pleurotus salmoneostramineus*. *Research Report of Rural Development Administration* (Korea) **31**, 48-54.

YOO, Y. B., SHIN, P. G., KIM, H. K., BYUN, M. O., YOU, C. H., CHA, D. Y. & CHANG, K. Y. (1989b). Transfer of isolated nuclei from *Agrocybe aegerita* mycelia into *Pleurotus florida* protoplasts. *Korean Journal of Mycology* **17**, 114-118.

YOO, Y. B., SONG, M. T., GO, S. J., YOU, C. H., CHA, D. Y., PARK, Y. H. & CHANG, K. Y. (1989c). Interorder protoplast fusion between *Pleurotus ostreatus* and *Ganoderma applanatum*. *Korean Journal of Mycology* **17**, 119-123.

YOU, C. H., YOO, Y. B., BYUN, M. O. & PARK, Y. H. (1988a). Studies on the transfer of isolated nuclei from *Pleurotus sapidus* into protoplasts of *Pleurotus ostreatus*. *Korean Journal of Mycology* **16**, 210-213.

YOU, C. H., YOO, Y. B. & PARK, Y. H. (1988b). Studies on protoplast formation and reversion of *Pleurotus sapidus* Kalchbr. *Korean Journal of Mycology* **16**, 214-219.

YU, M. Y. & CHANG, S. T. (1987). Effects of osmotic stabilizers on the activities of mycolytic enzymes used in fungal protoplast liberation. *MIRCEN Journal of Applied Microbiology and Biotechnology* **3**, 161-167.

CHAPTER 9

INTERSPECIFIC AND INTERGENERIC HYBRIDIZATION OF EDIBLE MUSHROOMS BY PROTOPLAST FUSION

Kihachiro Ogawa

Department of Biological Resource, Miyazaki University, Miyazaki 889-21, Japan.

1. INTRODUCTION

Breeding of edible mushrooms using the technique of spore-crossing is only applicable to intraspecific hybridization in monokaryotic strains. Thus, an alternative approach to the breeding of basidiomycetes involving protoplast fusion is now being widely adopted in order to obtain interspecific and intergeneric hybrids which produce fruit bodies of desirable quality. One of the principal advantages of protoplast fusion is hybridization between incompatible basidiomycetes. Identification of fusants obtained by protoplast fusion has generally been accomplished by nutritional complementation using auxotrophic strains. However, hybridization of auxotrophic mutants by protoplast fusion may lead to genetic imperfections. Therefore, we attempted to obtain new hybrids by using practical protoplast fusion techniques. Interspecific and intergeneric fusion products were obtained from protoplast fusion between basidiomycetes such as (1) *Pleurotus ostreatus* SM-1 [auxotrophic (*cytidine*$^-$) and monokaryotic strain] and *Pleurotus cornucopiae* TN-6 (monokaryotic strain), (2) *P. cornucopiae* TN3-1 [auxotrophic (*leucine*$^-$) and monokaryotic strain] and *Lentinus edodes* MYB-1 (monokaryotic strain), (3) *P. ostreatus* SM-A (monokaryotic strain) and *L. edodes* MYB-1, and (4) *P. cornucopiae* TN-6 and *Lyophyllum decastes* (Fr.) Sing..

There have been few occasions where fruit body formation was obtained with hybrids produced by interspecific fusion (Toyomasu & Mori, 1987; and Morinaga *et al.*, 1985) although intraspecific fusions of *Coprinus cinereus* (Toyomasu *et al.*, 1988), *P. ostreatus* (Ohmasa, 1986) and *Coprinus macrorhizus* (Kiguchi & Yanagi, 1985) have been relatively well studied.

We have obtained several hybrids from these basidiomycetes by protoplast fusion. These results show that improvement of these basidiomycetes can be easily accomplished using practical protoplast fusion. This technique has very important implications for the breeding of edible mushrooms.

2. PREPARATION OF PROTOPLASTS FROM MYCELIA OF BASIDIOMYCETES

A liquid medium (MYG) containing (1% malt extract, 0.4% yeast extract and 0.5% glucose in distilled water was used for mycelial cultivation. The same medium supplemented with 0.6 M mannitol was used as a regeneration complete medium (CM). A third medium containing (per litre): 0.5 g $MgSO_4$, 0.46 g KH_2PO_4, 1.0 g K_2HPO_4, 1.5 g $(NH_4)_2SO_4$, 120 μg thiamine, and 20 g glucose was used as a minimal medium (MM). All media were adjusted to pH 5.6 before autoclaving. For solid culture, agar was added at a concentration of 2% (w/v). MYG medium (20 ml in a 100 ml Erlenmeyer flask) was inoculated with a mycelial suspension from a previously grown MYG culture and incubated at 25°C for 6 days without shaking. The resultant mycelium (200 mg wet weight) was harvested by filtration, washed several times with sterile water, and incubated with 3 ml of lytic enzyme solution in a Monod shaking culture apparatus at 30°C for 1 to 2 hr. The lytic enzyme solution for *P. ostreatus*, *P. cornucopiae* and *L. decastes* strains contained 2% (w/v) new Usukizyme (Ogawa *et al.*, 1988) in RCM, pH 5.6. The lytic enzyme solution for *L. edodes* contained 2% Cellulase onozuka RS (Yakult Honsha Co., Ltd.) and 0.1% chitinase preparation (Sigma Co., Ltd.) in 0.05 M succinate buffer, pH 4.6, containing 0.6 M $MgSO_4$. After incubation, the reaction mixture was filtered through G2 and G3 porosity sintered glass filters, and the filtrate was centrifuged at 700 x g for 5 min. The sedimented protoplasts were suspended

TABLE 1. Mycolytic activities of various enzyme preparations.

Enzyme preparation	β-1, 3-Glucanase (units/ml)	Chitinase (units/ml)
Cellulase onozuka R-10 (Yakult Honsha Co., Ltd.)	7.78	0.007
Usukizyme* (Kyowa Kasei Co. Ltd.)	48.21	0.467
Funcelase* (Yakult Honsha Co. Ltd.)	46.26	0.478
Chitinase (Sigma Chemical Co.)	0.39	0.740

* Tannic acid-polyethylene glycol method (Toyama & Ogawa, 1975).

in 1 ml of MM (containing 0.6 M mannitol) which was osmotically balanced. The mycolytic activities of the enzyme sources used this study are shown in Table 1.

Young monokaryotic mycelia incubated for 6 days in stationary surface culture were employed for protoplast preparation. Although combinations of lytic enzyme preparations have generally been used for protoplast preparation (Abe *et al.*, 1982, 1984; Yanagi *et al.*, 1985), a mycolytic enzyme preparation derived from *Trichoderma*, viz. Usukizyme, was highly effective when used independently. Since this enzyme preparation is balanced in terms of β-1,3-glucanase, chitinase and protease activities, it did not require supplementation for protoplast preparation using *P. ostreatus*, *P. cornucopiae* and *L. decastes*. However, combined lytic enzyme preparations were necessary to produce *L. edodes* protoplasts (Fig. 1-A). MYG medium containing 0.6 M mannitol was used for protoplast regeneration (Fig. 1-B). Under these conditions, formation of the first hyphae from protoplasts of *P. ostreatus* as observed within 48 hr. Approximately 1.5% of the protoplasts underwent regeneration. These results showed that the rate of protoplast regeneration in this basidiomycete

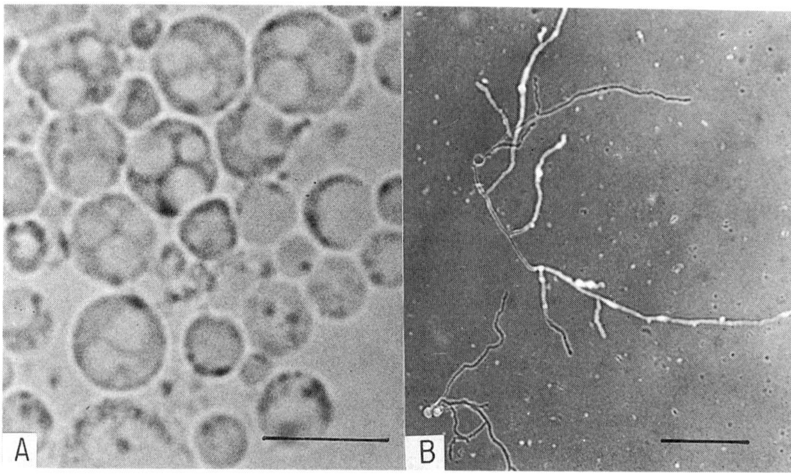

FIGURE 1. Protoplast regeneration in *P. ostreatus*.
A; Protoplasts from *P. ostreatus*.
B; Regeneration with the formation of the first hyphae after 2 days incubation. Protoplasts used in regeneration experiments were suspended in MYG medium containing 0.6 M mannitol. Bar markers represent 5 µm (A) and 100 µm (B).

was lower than that achieved with *Trichoderma reesei* and *Aspergillus awamori* var. *kawachi*.

3. HYBRIDIZATION BETWEEN BASIDIOMYCETES BY PROTOPLAST FUSION

3. 1. Hybridization of *P. ostreatus* and *P. cornucopiae*

A mixture of protoplasts from *P. ostreatus* SM-1 (*cytidine*$^-$) and *P. cornucopiae* TN-6 were treated with polyethylene glycol as a fusogen in order to obtain hybrids. Fusion was carried out according to a modification of the method described by Anne & Peberdy (1976). Mixed protoplasts were centrifuged at 700 x g for 5 min. Sedimented protoplasts were then resuspended in 1 ml of a solution containing 35% (w/v) polyethylene glycol 4,000 (PEG, average mol. wt., 3,000, Wako Pure Chem. Co., Ltd.) and 0.5mM $CaCl_2.2H_2O$ in

TABLE 2. Regeneration ratio after fusion of protoplasts prepared from monokaryotic strains of *P. ostreatus* and *P. cornucopiae*.

Protoplast pair	Protoplast After*	(CM) Control**	Colony (MM)	Regeneration (%)
P. ostreatus SM-1 (*cytidine*$^-$)				
	1×10^5	1.28×10^4	3.0×10^2	0.34
P. cornucopiae TN-6 (prototroph)				

Fusion was achieved by suspending the protoplasts in 35% PEG (4,000), 0.05 M $CaCl_2$, 0.05 M glycine-NaOH buffer, pH 7.5 (25°C for 15 min). After washing with 0.6 M NaCl, a series of appropriately diluted protoplast suspensions were plated onto hypertonic MM and CM agar.

* Protoplasts regenerated on hypertonic CM after fusion treatment.

** Colonies regenerated by the water-lysis control test (Ogawa *et al.*, 1988) after fusion treatment.

50mM glycine-NaOH buffer, pH 7.5. After standing at 25°C for 15 min., the suspension was centrifuged at 700 x g for 5 min. The sedimented protoplasts were resuspended in 1 ml of an osmotically balanced MM, plated onto the same medium containing 2% agar and 0.6 M mannitol, and then overlayed with MM containing 0.5% agar. Table 2 shows the regeneration ratio after fusion treatment of SM-1 and TN-6 protoplasts. The three hundred colonies which were obtained on MM agar consisted of both monokaryotic strains derived from TN-6 and fusion products. The fusion products were first selected on the basis of clamp connection formation and the barrage reaction of each colony (Kawasumi *et al.*, 1987) (Fig. 2). From the 300 colonies growing on MM agar, 10 colonies were selected as fusion products. These products showed apparent clamp connection formation in the mycelia. The observed frequency of dikaryon formation was 0.01%, or approximately 1000-fold higher than the reversion rate of *P. ostreatus* SM-1 (Table 3). As shown in Fig. 3, these dikaryons were capable of developing fruit bodies for long periods under standard cultivation conditions in a sawdust medium. The fruit bodies arising from these fusion products exhibited the yellowish coloration distinctive for *P. cornucopiae*, and the characteristic morphology

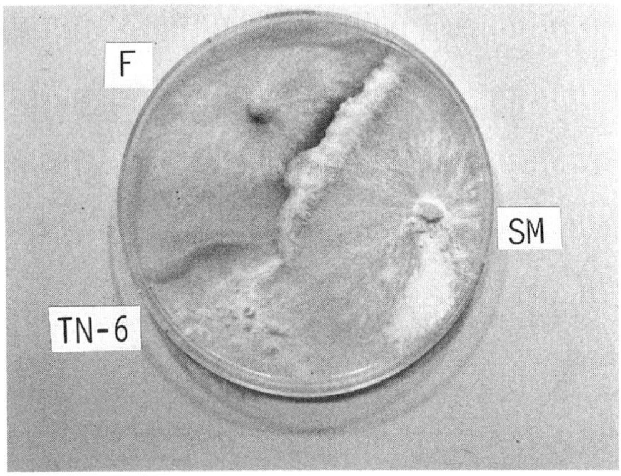

FIGURE 2. The dual culture of parental strains and a fusion product. F: fusion product; SM: *P. ostreatus* SM-1 (*cytidine* -) (auxotrophic and monokaryotic strain; TN-6: *P. cornucopiae* (prototrophic and monkaryotic strains).

TABLE 3. Reversion rate of auxotrophic mutants.

Strain	Reversion rate
P.ostreatus SM-1 (*cytidine*-)	< 1.2 x 10^{-7}
P. cornucopiae TN3-1 (*leucine*-)	< 1.7 x 10^{-7}

Reversion rate is defined as the ratio of number of protoplasts on MM and CM.

of *P.ostreatus*. Fruit body formation was carried out using the following procedure: 1 kg of solid culture medium consisting of 4 parts sawdust and 1 part rice bran was placed in a polyethylene bag and mixed with tap water to give a moisture content of 65%. After autoclaving at 120°C for 1 hr, the medium was inoculated with sawdust containing the spawn of the fusion product and kept at 15°C overnight. Excess water was then removed by decantation and the incubation continued at 15°C throughout the fruiting

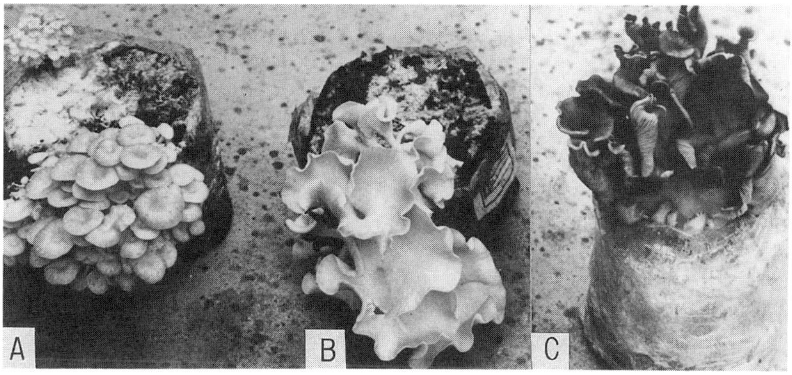

FIGURE 3. Fruit bodies of both parental strains and of a fusion product obtained following the fusion of protoplasts derived from *P. ostreatus* SM-1 and *P. cornucopiae* TN-6. A: *P. cornucopiae* (dikaryotic strain); B: fusion product, C; *P. ostreatus* (dikaryotic strain).

period.

3.2. Intergeneric Hybridization between Edible Mushrooms

Intergeneric fusion between protoplasts of *P. cornucopiae* (*leucine*$^-$) and *L. edodes* MYB-1 was attempted. The colonies obtained from regenerated protoplasts consisted of monokaryotic strains derived from MYB-1 and fusion products. Table 4 shows the regeneration ratio after fusion treatment of protoplasts from these two fungal strains. Fusion products were selected on the basis of the barrage reaction and clamp connection formation observed with each colony.

The frequency of dikaryon formation was 0.003%. In this case, fusion products formed fruit bodies which were of the same morphological type as those of *L. edodes* (Fig. 4).

Fusion products obtained from fusions between protoplasts derived from non-auxotrophic strains such as *P. ostreatus* SM-A and *L. edodes* MYB-1 or *P. cornucopiae* TN-6 and *L. decastes* (Fr.) Sing also produced fruit bodies under the conditions described above (Fig. 5 and 6). *L. edodes* and *L. decastes* grow very slowly compared with *P. ostreatus* and *P. cornucopiae* and protoplast fusion was carried out in order to produce faster growing hybrids.

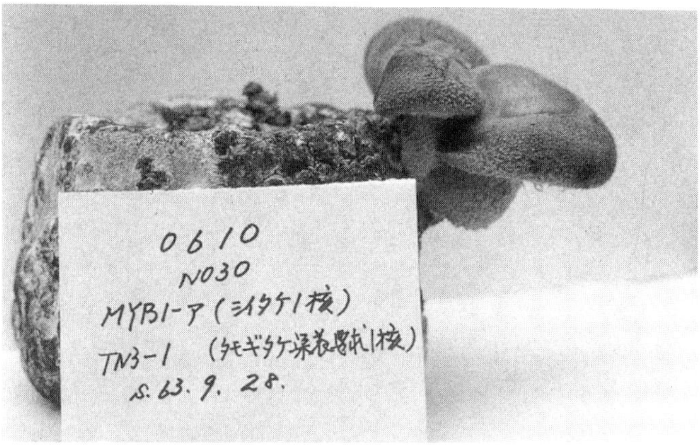

FIGURE 4. Fruit bodies of a fusion product obtained following the the fusion of protoplasts derived from *P. cornucopiae* TN3-1 (*Leucine*-) (auxotrophic and monokaryotic strain) and *L. edodes* MYB-1 (prototrophic and monokaryotic strain).

FIGURE 5. Fruit bodies of a fusion product obtained following the fusion of protoplasts derived from *P. ostreatus* SM-A (prototrophic and monokaryotic strain) and *L. edodes* MYB-1.

FIGURE 6. Fruit bodies of a fusion product obtained following the fusion of protoplasts derived from *P. cornucopiae* TN-6 and *L. decastes* (prototrophic and monokaryotic strain).

TABLE 4. Regeneration ratio after fusion of protoplasts prepared from monokaryotic strains of *P. cornucopiae* TN3-1 and *L. edodes* MYB-1.

Protoplast pair	Protoplast After*	(CM) Control**	Colony (MM)	Regeneration (%)
P. cornucopiae TN3-1 (*Leucine*⁻) / *L. edodes* MYB-1 (protoptroph)	$3.0x10^5$	$1.5x10^4$	$7.3x10^2$	0.26

* Protoplasts regenerated on hypertonic CM after fusion treatment.
** Colonies regenerated by the water-lysis control test after fusion treatment.
leucine⁻; leucine requiring.

Fusion products exhibited accelerated mycelial growth, and produced fruit bodies after only half the cultivation period required for *L. edodes* or *L. decastes* fruiting. None of the parental monokaryotic strains tested formed fruit bodies when cultured separately on the sawdust/bran medium under the conditions described.

Biochemical identification of the fusion products was established by analyzing for isozymes of esterase and malate dehydrogenase in parental strains and fusion products using polyacrylamide gel electrophoresis. Enzyme patterns of the fusion products were compared with those of parental strains in order to confirm interspecific and intergeneric hybridization. Figure 7 shows the malate dehydrogenase isozymes of fusion products obtained following the fusion of protoplasts derived from *P. cornucopiae* TN-6 and *L. decastes* (Fr.) Sing. The isozyme patterns of the fusant showed bands common to both parental strains. Similar results were also obtained in the case of the fusion products derived from other strains. These results confirmed that the fusion products arose as a result of interspecific and intergeneric hybridization.

The various stages in the breeding of basidiomycetes by protoplast fusion are represented in Fig. 8.

4. APPLICATIONS OF HYBRIDIZATION AND BREEDING OF BASIDIOMYCETES BY PROTOPLAST FUSION

Edible mushrooms are now attracting world-wide attention as nutritious and functional food-stuffs and as a sources of medicinal compounds. Since the fusion of protoplasts derived from basidiomycetes represents a highly feasible technique for interspecific and intergeneric hybridization, it provides a very useful tool for strain improvement among edible mushrooms.

Interspecific and intergeneric dikaryons were produced by PEG-induced fusion of protoplasts from *P. ostreatus* and *P. cornucopiae*, *P. ostreatus* and *L. edodes*, *P. cornucopiae* and *L. edodes*, and *P. cornucopiae* and *L. decastes*. The dikaryons were selected on the basis of semi- and non-nutritional complementation. The dikaryons obtained by protoplast fusion between basidiomycete species developed fruit bodies which exhibit morphologies intermediate between those of the original strains.

It is proposed that identification of hybrids generated by protoplast fusion can be readily achieved by 1) clamp connection formation, 2) barrage

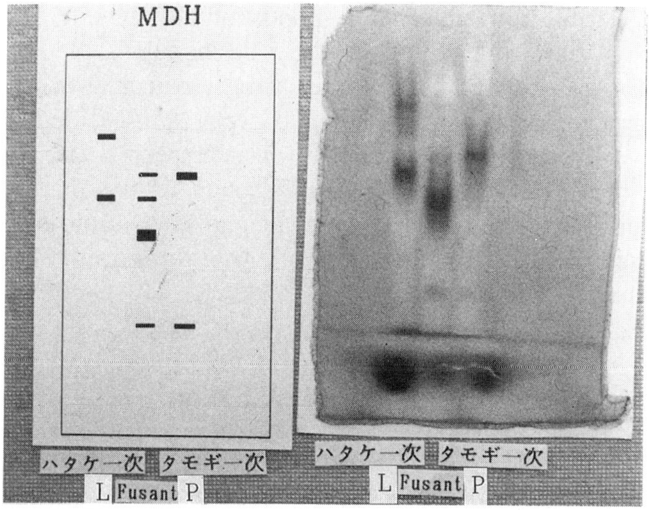

FIGURE 7. The isozyme pattern of malate dehydrogenase of a fusion product obtained following the fusion of protoplasts derived from *P. cornucopiae* TN-6 and *L. decastes*.

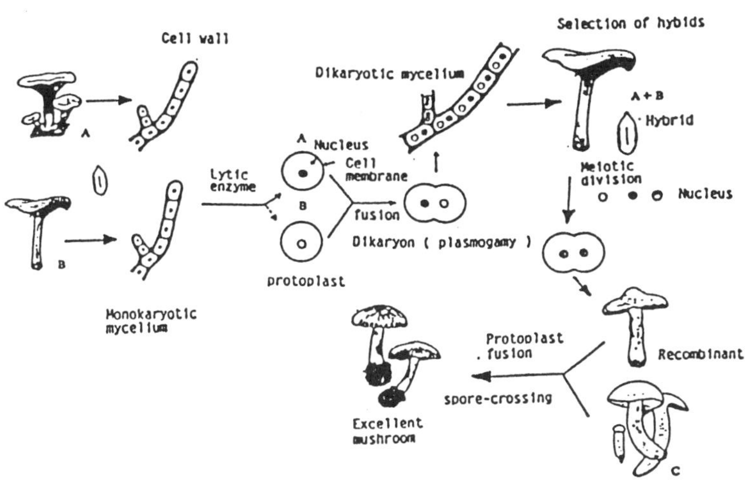

FIGURE 8. Breeding of Basidiomycete species by protoplast fusion.

reaction, 3) fruit body formation, 4) sporulation of fruit body, 5) analysis of isozyme patterns and 6) chromosomal analysis using pulse field electrophoresis.

As indicated by the data described here, hybridization using protoplast fusion has significance for 1) breeding between intraspecific incompatible basidiomycete species, 2) feasibility of interspecific and intergeneric hybridization, 3) development of multifunction such as the formation of physiological active substances (medicinal compounds), aromatic and growth acceleration, 4) growth acceleration of mycorrhiza mushrooms, and 5) genetic analysis.

REFERENCES

ABE, M., UMETSU, H., NAKAI, T. & SASAGE, D. (1982). Regeneration and fusion of mycelial protoplast of *Tricholoma matsutake*. *Agricultural and Biological Chemistry* **46**, 1955-1957.

ABE, M., UMETSU, H., NAKAI, T. & SASAGE, D. (1984). Regeneration of mycelial protoplasts from *Lyophyllum shimeji*. *Agricultural and Biological Chemistry* **48**, 1635-1636.

ANNE, J. & PEBERBY, J.F. (1976). Induced fusion of fungal protoplasts following treatment with polyethylene glycol. *Journal of General Microbiology* **92**, 413-417.

KAWASUMI, T., KIUCHI, N., FUTATSUGI, Y., OHBA, K. & YANAGI, S.O. (1987). High yield preparation of *Lentinus edodes* (Shiitake) protoplasts with regeneration capacity and mating type stability. *Agricultural and Biological Chemistry* **51**, 1649-1656.

KIGUCHI, T. & YANAGE, S.O. (1985). Intraspecific heterokaryon and fruit body formation in *Coprinus macrorhizus* by protoplast fusion of auxotrophic mutants. *Applied Microbiology and Biotechnology* **22**, 121-127.

MORINAGA, T., KIKUCHI, K. & MORI, R. (1985). Hybrid-fruit body formation from mushrooms by protoplast fusion. Abstracts of the Annual Meeting of the Society of Fermentation Technology of Japan, Osaka, p. 2.

OGAWA, K., BROWN, A.J. & WOOD, T.M. (1988). Intraspecific hybridization of *Trichoderma reesei* QM 9414 by protoplast fusion using color mutants. *Enzyme and Microbial Technology* **9**, 229-232.

OGAWA, K., TOYAMA, N. & SUGITA, K. (1988). Purification and some properties of β-1,3-glucanase from *Trichoderma viride*. *Hakkokogaku Kaishi* **66**, 385-391.

OHMASA, M. (1986). Intraspecific protoplast fusion of *Pleurotus ostreatus* using auxotrophic mutants. *Janpanese Journal of Breeding* **36**, 429-433.

TOYAMA, N. & OGAWA, K. (1975). Saccharification of agricultural cellulosic wastes with *Trichoderma viride* cellulase. *Symposium on Enzymatic Hydrolysis of Cellulose, Aulanko, Finland,* p. 375.

TOYOMASU, T., ARIMA, S. & MORI, K. (1988). Nuclear distribution of intraspecific protoplast fusants of *Coprinus cinereus*. *Transactions of the Mycological Society of Japan* **29**, 431-436.

TOYOMASU, T. & MORI, K. (1987). Fruit body formation of the fusion products obtained on interspecific protoplast fusion between *Pleurotus* species. *Agricultural and Biological Chemistry* **51**, 2037-2040.

YANAGI, S.O., MONMA, M., KAWASUMI, T., HONO, A., KITO, M. & TAKEBE, I. (1985). Conditions for isolation of the colony formation by mycelial protoplasts of *Coprinus macrorhizus*. *Agricultural and Biological Chemistry* **49**, 171-179.

CHAPTER 10

MOLECULAR TOOLS IN BREEDING *AGARICUS*

James B. Anderson

Department of Botany, University of Toronto,
Mississauga, Ontario L5L 1C6, Canada.

1. INTRODUCTION

Our interest has been to use molecular genetic markers to develop an effective strategy for breeding improved strains of *Agaricus bisporus* Lange (Imbach) (=*A. brunnescens* Peck). In this review, I will first summarize some recently developed molecular-genetic markers and then describe how we are using these markers to quantify genetic variability among strains and to follow the process of inheritance in *A. bisporus*. Before discussing markers, however, it is important to consider the sources of genetic variation in *A. bisporus* upon which a breeding program can be based.

1.1. Genetic Variation in *Agaricus bisporus*

It is now well established that cultivated strains of *A. bisporus* encompass only a small proportion of the total genetic variability in the species (Castle *et al.*, 1987; Kerrigan & Ross, 1989; Royse & May, 1982). For example, in recent years, a substantial proportion of the production of *A. bisporus* in the West is from one enormously productive strain, the UI developed by Dr. Gerda

Fritsche at the Horst Station in the Netherlands (Fritsche, 1991), and its derivatives. Other strains of *A. bisporus* are now prevalent in China (Wang *et al.*, 1991). Given the limited variability among existing cultivars, it is obvious that additional genetic variability for a variety of traits is desirable, if not essential, for meeting the many conditions and problems likely to arise in the mushroom growing industry. Clearly most of this variability must come from wild-collected strains. Wild variants, including specific genes and clusters of genes, are the result of a long period of evolution, including mutation, recombination, and natural selection. These natural variants have stood the test of time. It is unlikely that we can recover the full range of variation needed by mutagenesis and selection in the laboratory, within a reasonable period of time. Similarly, in the near future, DNA based transformation methods (Fincham, 1989) now under development in *A. bisporus* cannot be expected to generate superior commercial strains. Of course, transformation will be extremely important as a tool in basic studies of genes and their regulation in *Agaricus*.

If wild populations must serve as the primary source of new genetic variability for the mushroom industry, it is vital that extensive collection efforts be made immediately as wild populations are threatened in several ways. First, climate change might reduce the longevity of the vegetative mycelium and affect patterns of fruiting and colonization. Second, indigenous habitats harboring novel genotypes of *A. bisporus* are rapidly being destroyed. Third, the industry has selected a few genotypes from the wild and propagated them extensively in cultivation. Unlike most crop plants, cultivated strains of *A. bisporus* have been changed very little; these strains are fully capable of escaping and surviving in the wild. The mass escape of cultivar genotypes may well change existing gene frequencies by dilution, the ultimate result of which will be the impoverishment of natural gene pools. While it is impossible to objectively assess the seriousness of any of these threats to genetic diversity in wild populations of *A. bisporus*, there is strong evidence that other mushroom species are either less frequent or extinct in certain locations (Arnolds, 1991). The prudent course is to collect as many strains of *A. bisporus* from the wild as possible and to store these cultures with the best methods available.

2. GENETIC MARKERS

Following collection of *A. bisporus* strains from the wild, we may now consider the markers that can be used as reference points in the genome for characterizing wild material and for tracking the inheritance of determinants of important traits such as color, temperature optima, resistance to disease, or any other genetically determined trait of interest to the grower. The genetic markers themselves, however, serve mainly as neutral reference points in the genome and are in most cases unrelated to the phenotype of the strain. The ultimate objective is to find clearly recognized markers that show variability among strains and that are closely linked genetically to the determinants of important traits. In this way, predictions about the frequency and range of phenotypes arising from particular crosses can be made.

2.1. Auxotrophic Markers

The first unambiguous markers were auxotrophic mutations used by Raper *et al.* (1972) and Elliott (1972, 1985) in establishing the basic facts of the life cycle of *A. bisporus*. Auxotrophic strains, most of which were produced by mutagenic treatment in the laboratory, carry recessive genetic lesions that result in the loss of a biosynthetic capability. The resulting phenotype is a new nutritional requirement such as for a particular amino acid, nitrogenous base, or vitamin. The auxotrophic phenotype is easy to score as the failure to grow on a minimal medium containing defined carbon and nitrogen sources, several inorganic salts, trace elements, and, in *A. bisporus* and many other higher fungi, the vitamin thiamin. The auxotrophic requirement(s) can be determined by supplementing minimal medium with the substances in question and asking which supplement restores the ability of an auxotrophic strain to grow.

An important characteristic of auxotrophic markers is that nuclei carrying different requirements will show complementation when combined in a heterokaryon; in such a cell the deficiency of one nucleus is corrected by the corresponding wild-type gene carried by the other nucleus. If the objective is to use complementation for a selective screen, for example in identifying hybrids produced by hyphal anastomosis or protoplast fusion, then auxotrophs will continue to serve an important function. Unfortunately, auxotrophs are very difficult to recover. This is primarily because (a) auxotrophic mutations are recessive and (b) the spores and hyphae of *A. bisporus* are multinucleate. Another possible problem is that many DNA sequences appear to be present in more than one copy in the genome of *A. bisporus* (Anderson, J.B., unpublished observation). The number of auxotrophs currently available is therefore very small. Even if more auxotrophs were available, however, another problem would need to be solved: the mutagenic process used to induce auxotrophic mutations may result in unintended genetic damage to the strain.

2.2. Allozyme Markers

More recently, genetic markers that are naturally occurring and presumably have little or no effect on the appearance or behavior of the strain have been developed in *A. bisporus* (Royse & May, 1982; Kerrigan & Ross, 1989; Wang *et al.*, 1991). Allozymes are proteins in which amino acid substitution(s) result in a change in electrophoretic mobility due to a change in the net charge or the conformation of the protein. After electrophoresis, the allozymes are identified by specific staining for the particular enzymatic activity. As a result, the allelic forms of a given enzyme can be identified. When properly interpreted, allozyme loci behave in a Mendelian manner in crosses. This is because the amino acid substitutions in allozymes are encoded in the DNA, the primary genetic material. Under controlled conditions, allozymes are powerful genetic markers. They are relatively inexpensive to use and the return of data per unit of effort at the laboratory bench is very high. Unfortunately, there are only about ten allozyme markers available in *A. bisporus*.

2.3. DNA Based Markers

If the objective is to find neutral markers in the primary genetic material, a practically inexhaustible source is available when variation in the DNA itself is examined. For most breeding purposes, it does not matter whether the DNA region showing variability is found within a gene or between genes. Two kinds of DNA based markers are particularly relevant to the genetics of *A. bisporus*: restriction fragment length polymorphisms (RFLPs) and random amplified polymorphic DNAs (RAPDs). For RFLP or RAPD analysis, a variety of reliable methods for extracting fungal DNA (Taylor & Natvig, 1987) can be used.

2.3.1. RFLPs. RFLPs are based on variation in cleavage patterns produced by bacterial restriction endonucleases. These enzymes recognize specific short sequences in the DNA and make a cut at, or near, that site. Restriction enzymes are part of the defense systems of bacteria, which are relatively "leaky" to invading DNA, as compared with eukaryotes. Foreign DNA is quickly fragmented by the endogenous restriction endonuclease after entering a bacterial cell. At the same time, bacteria have parallel modification systems that effectively make restriction sites in their own DNA unavailable for digestion. Many different Type II restriction enzymes of varying specificities are commercially available.

Southern hybridizations (Southern, 1975) are most often used to detect the specific restriction fragments among the many thousands of fragments found in the genome of an organism. For example, the haploid genome size of *A. bisporus* is about 34,000,000 base pairs (Arthur *et al.*, 1982). If the average size of an EcoRI fragment is about 4,000 base pairs, then about 8-9,000 fragments are expected to be generated by cleavage with EcoRI, far too many to resolve as separate bands on one agarose gel. In the Southern hybridization procedure, the genomic DNA is cut with a restriction enzyme. The fragments produced by specific cleavage are then separated according to size by agarose gel electrophoresis. The DNA in the gel is made single-stranded by alkali treatment, and is transferred to a hybridization membrane, usually nitrocellulose or nylon. A small segment of homologous DNA, usually less than 10,000 bp cloned in a bacterial plasmid, is used as a probe. The probe DNA is made radioactive and allowed to hybridize with the filter-bound DNAs. After washing, fragments of genomic DNA hybridizing to the radioactive probe are detected by autoradiography. Recently, several nonradioactive methods for labelling probe DNAs have been developed.

Polymorphisms among individual haplotypes are often evident in the pattern of hybridizing fragments. The variation in restriction fragment sizes can arise from essentially two kinds of variation in the target DNA. A strain may carry specific base substitutions that either create or destroy a restriction-enzyme recognition site in the target DNA region, relative to other strains. Also, rearrangements, such as insertions or deletions, may alter the sizes of restriction fragments in the target region, as well as the number of recognition enzyme cleavage sites present. Figure 1 shows some of these possibilities. Once restriction polymorphisms are identified, the various banding phenotypes can be interpreted as codominant alleles of specific loci in much the same way

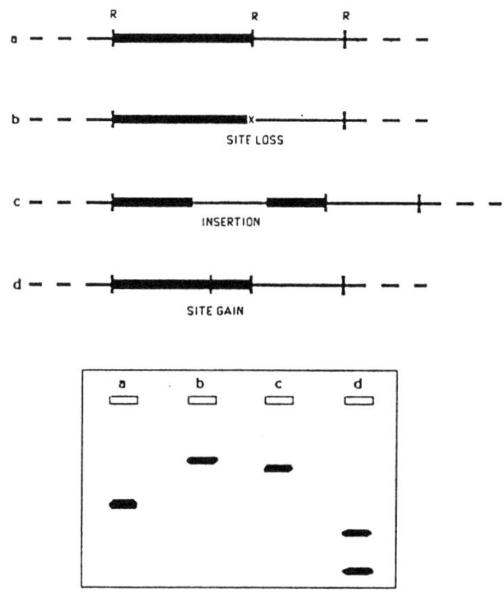

FIGURE 1. Restriction polymorphisms in a specific genomic region of four haploid individuals (top) and the corresponding phenotypes on Southern blots (bottom) probed with the reference sequence (thick line).

as for allozyme markers. RFLPs can then be used as Mendelian markers for genetic mapping (Botstein *et al.*, 1980; Hulbert *et al.*, 1988; Chang *et al.*, 1988).

2.3.2. RAPDs. Another class of markers is based on a modified form of the polymerase chain reaction (PCR) used to amplify a specific region of genomic DNA *in vitro* (Innis *et al.*, 1990). In the standard form of PCR, some sequence information for the region to be amplified is required. Chemically synthesized, single-stranded primers, usually 15-30 bases in length, anneal to specific sites flanking the target region of genomic DNA. Each primer initiates DNA synthesis from its 3' end. PCR reaction mixtures contain the target DNA, the two primers, the four deoxynucleoside triphosphates, and a thermostable DNA polymerase plus Mg^{++}, and a buffer. PCR mixtures are then subjected to cycles of denaturation at 94° C, primer annealing at 40-70° C, and DNA polymerization at 72°C. This temperature cycling results in an exponential increase of the DNA segment located between the 5' ends of the primers. One of the breakthroughs leading to the widespread use of PCR was the development of the thermostable DNA polymerase, the activity of which is not destroyed during the denaturation step. The automated technology to control temperature has also made the entire process more practical. Finally, the products of PCR can be restriction mapped, cloned, or sequenced.

Random Amplified Polymorphic DNA (RAPD) markers (Williams *et al.*, 1990; see also Welsh and McClelland, 1990) result from a variant type of PCR using shorter primers of arbitrary sequence, usually 10 bases in size and 50-80% G+C. (Some of our best primers were designated by tossing coins, six of which had "G" written on one side and "C" written on the other and four of which had "A" on one side and "T" on the other.) The rationale for the RAPD procedure is that within the genome, there is a certain distribution of annealing sites for any single, arbitrary primer. This distribution is thought to result mostly from chance. Some priming sites are close enough together, and in the inverted orientation, to yield an amplified product in a PCR reaction in which the annealing temperature is typically low (36°C). Such RAPD products are polymorphic among individuals for much the same reasons as for variability in restriction fragment sizes among individuals (Figure 2). The advantage of RAPD analysis is that it is simple and fast. The RAPD procedure requires few materials, and the amplification products can be visualized on agarose gels with no Southern hybridization and no radioactive materials. The disadvantage of RAPD analysis is that reactions can be extremely sensitive to conditions such as the concentrations of reaction components, contaminants in the target DNA preparation, enzyme source, and especially the temperature cycling profile. Great care must be taken to show that the RAPD markers are repeatable in fully independent experiments. Given proper controls, however, the RAPD technique can be extremely powerful.

3. GENETIC VARIABILITY IN POPULATIONS

With a set of markers available, questions can be asked about the distribution of genetic variability in cultivated strains and in wild populations (Royse & May, 1982; Kerrigan & Ross, 1989). For example, we recently asked whether the isolates found within one habitat are on average more similar to one

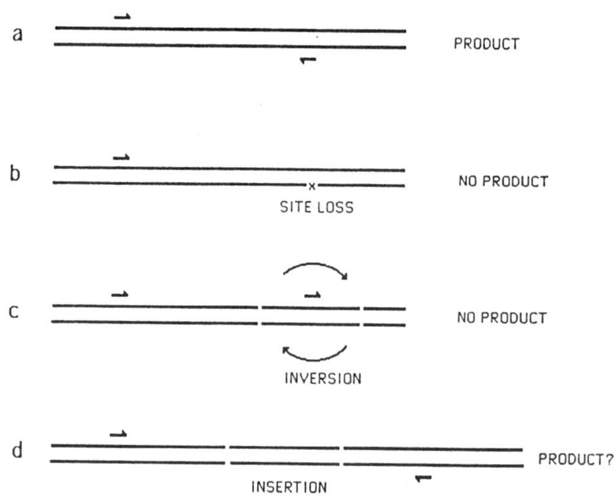

FIGURE 2. Polymorphisms in a genomic region of four haploid individuals that result in the loss of a Random Amplified Polymorphic DNA product.

another than to the population at large (Kerrigan, R.W., Horgen, P.A. & Anderson, J.B., submitted for publication). This recent study used 29 heterokaryotic strains collected from various locations in California and made comparisons between the wild-collected and cultivated heterokaryons. In total, we assayed 21 different genetic loci, 7 allozyme and 14 RFLP, and scored the alleles carried by each heterokaryotic strain at each locus. Then, we calculated genetic distance values for each pairwise combination of isolates. Lastly, we used the FITCH-MARGOLIASH clustering algorithm to represent the distance relationships in the form of a cluster diagram. The results are shown in Figure 3.

Essentially there were three clusters of isolates. The only strain from a dry area of California (central valley, near Davis), K1421, was very different from all others and appeared in a cluster by itself. Clearly, it will be interesting to collect more strains from this area. Another cluster of isolates was from a distinct woodland habitat, coastal cypress groves. The remaining strains were from disturbed areas (golf courses, lawns) as well as from coastal cypress groves, and formed a large and heterogeneous cluster. These strains show a wide range of similarity values to one another, and some of the strains were genetically very similar to existing cultivated types.

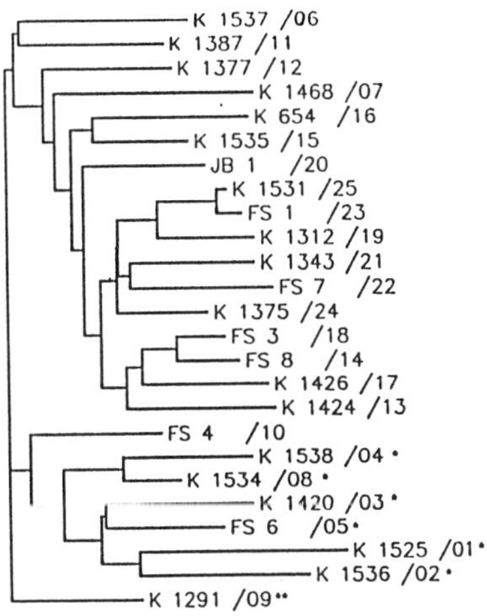

FIGURE 3. Phenogram of pairwise genotypic distances produced by one of ten FITCH evaluations of the input data matrix. The topology of this tree is the same as the consensus of ten FITCH evaluations performed separately. Horizontal branch length is proportional to genotypic distance; vertical lines serve only to separate horizontal branches. The tips of branches are marked with only one strain representing the particular genotype. A single asterisk(*) denotes isolates representing the putatively "indigenous" California genotype from the cypress habitat. A double asterisk(**) denotes the single isolate collected from a dry region of California.

The simplest interpretation of the genetic distance data is that the California population is composed of at least two ancestral types, one indigenous to cypress groves and the other introduced to disturbed areas as escapes from cultivation. Subsequently, hybridization occurred between the two ancestral types, and alleles common in cultivated strains are now also found in the cypress habitat. Of course, other more complex explanations of ancestry are plausible. Clearly, however, existing genetic variability in the California population is at least partially partitioned by habitat.

Further evidence for hybridization of the two ancestral types was that mitochondrial DNA types, visualized by probing restriction digests of genomic DNA with cloned segments of mitochondrial DNA, were of limited diversity among the cultivar-like isolates but were of extensive diversity among the putatively indigenous types (Kerrigan, R.W., Horgen, P.A. &

Anderson, J.B., submitted for publication). The putative hybrids had mixtures of nuclear alleles found only in the indigeous types and nuclear alleles common in cultivated strains, but had a variety of mitochondrial types that were very different from those commonly found in cultivated strains .

The question now arises as to whether the cluster of indigenous isolates from cypress cohere as a distinct group when compared to a worldwide set of collections. Preliminary analysis of isolates from Alberta and Ontario, Canada, as well as Israel, indicate that the California cypress group does indeed remain distinct (Kerrigan, R.W., Horgen, P.A. & Anderson, J.B., unpublished results). This preliminary result leads us to ask how many other indigenous populations remain to be discovered and emphasizes the urgency of sampling wild *A. bisporus* more extensively on a worldwide scale. Our results to date further suggest that we must know more about the structure of natural populations, especially with regard to habitat type, if we are to maximize the recovery of genetic variability in wild-collected material.

4. CROSSING

Once collections encompassing a range of genetic variability are characterized for their fruiting and production characteristics, it will be necessary to conduct crosses for long term breeding and improvement. The main obstacle to crossing has been the predominantly secondarily homothallic life cycle of *A. bisporus*; most basidiospores receive two of the four post meiotic nuclei and germinate to produce fertile heterokaryons. For crossing it is necessary to obtain homokaryons that are self-sterile until mated with another homokaryon. Homokaryons can be obtained from two sources: the rare homokaryotic basidiospores receiving only a single post meiotic nucleus (Raper *et al.*, 1972; Elliott, 1972, 1985; May & Royse, 1982) and the rare homokaryotic protoplasts from vegetative heterokaryons (Castle *et al.*, 1987). The two types of homokaryons are genetically very different. The former is the product of meiosis and is recombinant throughout much of its genome, while the latter is the product of mitosis and is completely non-recombinant, or very nearly so. Once homokaryons are obtained, crosses can be made to produce new hybrid heterokaryons (Castle *et al.*, 1988; May & Royse, 1982), most of which are fertile. When obtaining homokaryons and recovering hybrids it is desirable to confirm each step of the process with genetic markers.

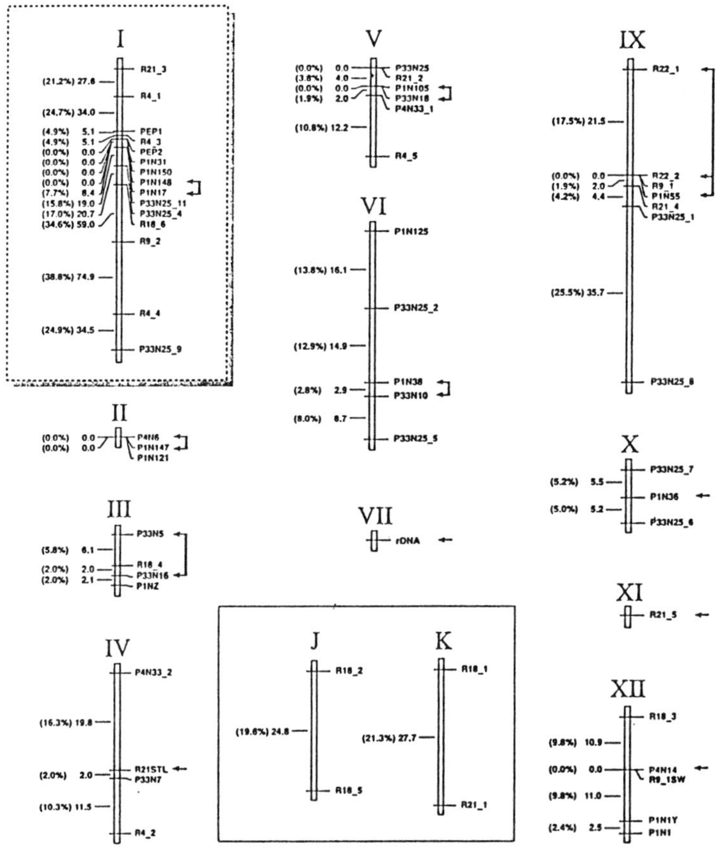

FIGURE 4. Genetic map of the 57 nuclear markers showing genetic linkage in *A. bisporus* cross AG 93b. Roman numerals refer to specific chromosome-sized DNAs separated by pulsed field gel electrophoresis; arrows represent marker DNAs used as probes to establish correspondence between genetic linkage groups and chromosomal DNAs. The map for chromosome 1 is drawn at one-fifth scale relative to the others. Linkage groups J and K are based solely on RAPD markers and could not be assigned to specific chromosomal DNAs. The amplification product of R21_5 and cloned rDNA hybridized to chromosomal DNAs VII and XI, respectively, both of which were genetically unmarked. The two additional genetically unmarked chromosomal DNAs VIII and XIII are not shown in this figure. Marker p33n25_3 (not shown) cosegregated with p33n10 on chromosome VI .

5. GENETIC MAPPING

With the basic process of crossing confirmed, we wished to characterize the general features of meiosis and the overall structure of the genome in *A. bisporus*. We therefore followed the patterns of transmission of 64 segregating genetic markers, including allozymes, RFLPs, and RAPDs, in 52 effectively homokaryotic offspring of a marked cross (strains Ag1-1 and Ag89-65) of *A. bisporus* (Kerrigan, R.W., Horgen, P.A. & Anderson, J.B., submitted for publication). In all, 57 markers showed clear-cut linkage relationships, while seven markers showed complex or ambiguous linkage relationships (Figure 4). Departures from Mendelian segregation ratios were found in about 20% of the marker loci and in several cases were associated with determinants of homokaryon vigor, especially in the case of Linkage group III in which the Ag89-65 parent carries a recessive gene or genes that produce slow growth. Despite the skewed segregation of some markers, both joint and independent segregation of markers were observed. Chi-square tests of independent segregation ($P<0.01$) for all pairwise combinations of markers identified nine distinct linkage groups, each of which corresponded to a different chromosome-sized DNA resolved by pulsed-field electrophoresis (Royer, J.C., Hintz, W.E., Kerrigan, R.W. & Horgen, P.A., submitted for publication). The order of markers within each linkage group and their optimized map intervals (Haldane mapping function) were determined with the maximum likelihood procedures found in the computer program MAPMAKER (Lander *et al.*, 1987). One anonymous DNA marker (R21_5) was physically localized to a tenth chromosome-sized DNA and the rDNA repeat was found on an eleventh. Two remaining genetic linkage groups marked only by RAPDs in regions that are repeated in the genome may or may not correspond to the two remaining chromosome-sized DNAs observed; establishing a complete correspondence of genetic and physical linkage will require additional markers in single-copy DNA.

Although a limited number of genetic markers have been previously mapped (Spear *et al.*, 1983; Wang *et al.*, 1991), the present genetic map study is, to our knowledge, the first to examine inheritance patterns in this commercially important mushroom with a set of confirmed homokaryons and a set of markers sufficient to follow the majority, if not all, chromosomes through meiosis. Our data show that, despite its secondarily homothallic life history, *A. bisporus* has an essentially conventional meiosis in which both independent assortment and joint segregation of markers occur, but crossing over is infrequent over much of the mapped genome. This information will have a direct bearing on the strategies for obtaining progeny with the desired combinations of traits from artificial crosses.

Our remaining genetic mapping objectives are first to examine additional crosses. The question here is how consistent or variable is the genetic map among different heterokaryons. Already our data suggest chromosomal differences among the cross recently analyzed (Kerrigan, R.W., Horgen, P.A. & Anderson, J.B., submitted for publication). We also plan to determine the map position of the mating-type locus, provided the two mating types segregating from each cross can be scored unambiguously. Lastly, to make the genetic maps practical in a breeding program, it will be most important to locate the determinants of phenotypic traits on this map. This will be an expensive and labor-intensive task that is likely to be done in industrial research units or in government institutes for proprietary strain development.

6. BREEDING STRATEGIES

The final part of this review addresses how we can apply the genetic information to the problem of selection and breeding of improved strains. In all aspects of the process of strain improvement, it is important to separate the two factors controlling the performance of a particular strain: the environment and the genetic constitution. To identify genetic variability, the environmental regime must be held constant. But which set of environmental conditions is "right"? It may be important to identify more than one environmental regime appropriate to various local needs. Testing strains in a range of environments is desirable because a strain performing poorly under one environmental regime may perform extremely well under another. Defining and controlling the fruiting environment, however, is perhaps the most difficult and expensive part of strain improvement. Once suitable environmental regimes have been determined, the following is a series of approaches that can be applied, consistent with available resources, to the problem of finding improved strains of *A. bisporus*.

6.1. Testing Genetically Diverse Wild-Collected Strains

If a large number of strains are screened for fruiting characteristics, it may well be that particular strains will go a long way toward fulfilling local needs without any genetic modification. This kind of screening, which might yet prove to be the most cost-effective, can be accomplished even without access to molecular-genetic markers.

6.2. New Hybrids By Shuffling Intact Haplotypes

In this crossing protocol, mitotically derived homokaryons are identified among the protoplast regenerates obtained from heterokaryons and are then mated in many different pairwise combinations. Each of the haploid genotypes remains intact with respect to the original heterokaryons. This type of crossing can be done on a large scale and the new hybrids can be screened for the variability in fruiting characteristics, relative to the parent strains. Our own preliminary observations suggest that much phenotypic variability can be released by this method (Anderson, J.B., unpublished observations).

6.3. New Hybrids By Shuffling Chromosomes and Segments of Chromosomes

In this phase of a strain improvement program, additional new hybrids are obtained from the promising combinations identified in the previous phase of the breeding program. These crosses, however, are made with meiotically derived homokaryons identified among single basidiospore isolates. Each of these meiotically derived genotypes shows independent assortment of linkage groups and, less frequently, crossing over within linkage groups. If the phenotypic determinants of interest are on different chromosomes, then only relatively small numbers of homokaryons need to be tested to find the desired recombinants. If the phenotypic determinants are on the same chromosome, then substantially larger numbers of meiotically derived homokaryons may need to be tested. The size of the chromosomal interval within which recombination is desired will be inversely related to the number of homokaryons that must be screened.

Earlier breeding using meiotically derived homokaryons has already produced dramatic gains in quantity and quality of mushroom production (Fritsche, 1991). It seems reasonable that the inclusion of a wider base of germplasm, with molecular genetic markers used in a crossing program, can be expected to produce many further gains in the cultivation of *A. bisporus* worldwide.

REFERENCES

ARNOLDS, E. (1991). Mycologists and nature conservation. In *Frontiers in Mycology*, pp. 243-264. Edited by D.L. Hawksworth. United Kingdom: C. A. B. International.

ARTHUR, P., HERR, F., STRAUS, N., ANDERSON, J. & HORGEN, P. (1982). Characterization of the genome of the cultivated mushroom, *Agaricus brunnescens*. *Experimental Mycology* **7**, 127-132.

BOTSTEIN, D., WHITE, R., SKOLNICK, M. & DAVIS, R.W. (1980). Construction of a genetic linkage map in man using restriction fragment length polymorphisms. *American Journal of Human Genetics* **32**, 314-331.

CASTLE, A.J., HORGEN, P.A. & ANDERSON, J.B. (1987). Restriction fragment length polymorphisms in the mushrooms *Agaricus brunnescens* and *Agaricus bitorquis*. *Applied and Environmental Microbiology* **53,** 816-822.

CASTLE, A.J., HORGEN, P.A. & ANDERSON, J.B. (1988). Crosses among homokaryons from commercial and wild-collected strains of the mushroom *Agaricus brunnescens* (=*A. bisporus*). *Applied and Environmental Microbiology* **54,** 1643-1648.

CHANG, C., BOWMAN, J.L., DEJOHN, A.W., LANDER, E.S. & MEYEROWITZ, E. (1988). Restriction fragment length polymorphism linkage map for *Arabidopsis thaliana*. *Proceedings of the National Academy Sciences of the United States of America* **85**, 6865-6868.

ELLIOTT, T.J. (1972). Sex and the single spore. *Mushroom Science* **8,** 11-18.

ELLIOTT, T.J. (1985). The genetics and breeding of species of *Agaricus*. In *The Biology and Technology of the Cultivated Mushroom*, pp. 111-129. Edited by P. B. Flegg, D. M. Spencer & D. A. Wood. New York: John Wiley & Sons, Ltd.

FINCHAM, J.R.S. (1989). Transformation in fungi. *Microbiology Reviews* **53**, 148-170.

FRITSCHE, G. (1991). A personal view on mushroom breeding from 1957

- 1991. In: *Genetics and Breeding of Agaricus*, pp. 3-18. Edited by. L.J.L.D. van Griensven. Wageningen, The Netherlands: Centre for Agricultural Publishing and Documentation.

HULBERT, S.H., ILOTT, T.W., LEGG, E. J., LINCOLN, S.E., LANDER, E. S. & MICHELMAORE, R. W. (1988). Genetic analysis of the fungus *Bremia lactucae* using restriction fragment length polymorphisms. *Genetics* **120**, 947-958.

INNIS, M.A., GELFAND, D.H., SNINSKY, J.J. & WHITE, T.J. (Editors). (1990). *PCR Protocols: A Guide to Methods and Applications*. SanDiego: Academic Press.

KERRIGAN, R.W. & ROSS, I.K. (1988). Extracellular laccases: biochemical markers for *Agaricus* systematics. *Mycologia* **80,** 689-695.

KERRIGAN, R.W. & ROSS, I.K. (1989). Allozymes of a wild *Agaricus bisporus* population: new alleles, new genotypes. *Mycologia* **81**, 433-443.

LANDER, E.S., GREEN, P., ABRAHAMSON, J., BARLOW, A., DALY, M.J., LINCOLN, S.E. & NEWBURG, L. (1987). MAPMAKER: An interactive package for constructing primary genetic linkage maps of experimental and natural populations. *Genomics* **1**, 174-181.

MAY, B. & ROYSE, D.J. (1982). Confirmation of crosses between lines of *Agaricus brunnescens* by isozyme analysis. *Experimental Mycology* **6**, 283-292.

RAPER, C.A., RAPER, J.R. & MILLER, R.E. (1972). Genetic analysis of the life cycle of *Agaricus bisporus*. *Mycologia* **64,** 1088-1117.

ROYSE, D.J., & MAY, B. (1982). Use of isozyme variation to identify genotypic classes of *Agaricus brunnescens*. *Mycologia* **74,** 93-102.

SOUTHERN, E. M. (1975). Detection of specific sequences among DNA fragments separated by gel electrophoresis. *Journal of Molecular Biology* **98**, 503-517.

SPEAR, M.C., ROYSE, D.J. & MAY, B. (1983). A typical meiosis and joint segregation of biochemical loci in *Agaricus brunnescens*. Journal of *Heredity* **74,** 417-420.

TAYLOR, J. & NATVIG, D.O. (1987). Isolation of fungal DNA. In *Zoosporic Fungi in Teaching and Research*, pp. 252-258. Edited by M.S. Fuller & A. Jaworski. Athens, Georgia: Southeastern Publishing Corporation.

WANG, Z.S., LIAO, J.H., LI, F.G. & WANG, H.C. (1991). Studies on genetic basis of esterase isozyme loci Est A, B, and C in *Agaricus bisporus*. *Mushroom Science* **13,** 3-9.

WELSH, J. & McCLELLAND, M. (1990). Fingerprinting genomes using PCR with arbitrary primers. *Nucleic Acids Research* **18**, 7213-7218.

WILLIAMS, J.G.K., KUBELIK, A.R., LIVAK, K.J., RAFALSKI, J.A. & TINGEY, S.V. (1990). DNA polymorphisms amplified by arbitrary primers are useful as genetic markers. *Nucleic Acids Research* **18**, 6531-6535.

CHAPTER 11

MULTILOCUS ENZYME ELECTROPHORESIS FOR THE GENETIC ANALYSIS OF EDIBLE MUSHROOMS

Daniel J. Royse and Bernie May

Department of Plant Pathology, The Pennsylvania State University, University Park, Pennsylvania 16802, U.S.A.

1. INTRODUCTION

Allozymes are enzymes differing in electrophoretic mobility as the result of allelic differences in a single gene. Allozyme analysis allows informative data sets because the phenotypic differences in electrophoretic banding patterns between individuals can be directly correlated to genotypic differences. Thus, allozyme electrophoresis provides unambiguous codominant genotypes that can be used successfully in conducting genetic studies of fungi.

While the study of allozyme variation has been a standard tool in both eucaryotic and procaryotic genetic studies for more than two decades (Harris, 1966; Lewontin & Hubby, 1966; Selander *et al.*, 1986), mushroom scientists have only recently used the methodology. Within the last ten to twelve years, several species of edible mushrooms (Table 1) have been examined by allozyme electrophoresis to estimate genetic diversity and divergence, confirm crosses, identify homokaryotic breeding stocks, identify genotypic classes and map linkage groups in five mushroom genera. Extensive allozyme variation has been found in virtually all edible fungi studied. In this paper we review the successful applications of allozyme data to the genetics of edible

TABLE 1. Genera and common name of edible mushrooms to which multilocus enzyme electrophoresis (MEE) has been applied.

Species	Mushroom Common Name
Agaricus bisporus	Button
Agaricus campestris	Meadow
Lentinula edodes	Shiitake
Morchella spp.	Morel
Pleurotus spp.	Oyster
Volvariella volvacea	Paddy straw

mushrooms in hopes that other mushroom scientists may see the value of this methodology.

2. HISTORICAL PERSPECTIVES FOR EDIBLE MUSHROOMS

Allozyme electrophoresis was first used to elucidate the genetic life histories of edible mushrooms in the early 1980s (May & Royse, 1981). Prior to 1981, electrophoretic studies of edible mushrooms were confined to general protein patterns, utilizing some specific enzymes (Shechter *et al.*, 1973; Mouches *et al.*, 1979; Raper & Kaye, 1978; Paranjpe *et al.*, 1979). The electrophoretic phenotypes were evaluated on the presence or absence of particular bands without establishment of the genetic bases of these bands as did some earlier fungal studies (Speith, 1975). Thus, earlier studies were unable to take advantage of the one-to-one correlation of electrophoretic phenotype to individual genotype.

In a series of papers on edible fungi, we have shown the applicability of allozyme data to analyses of genetic relatedness and to practical breeding strategies in edible fungi (Bowden *et al.*, 1991a; May & Royse, 1981, 1982a, b, 1988; May *et al.*, 1988; Royse & May, 1982a, b, 1987; Royse *et al.*, 1983a, b; Royse *et al.*, 1987; Spear *et al.*, 1983). This groundwork was used later to identify new alleles and new genotypes in *A. bisporus* (Kerrigan & Ross, 1989) and in *Pleurotus* spp. (Kulkarni *et al.*, 1987) and to provide evidence of genetic divergence in *A. bisporus* (Kerrigan, 1990).

3. RESULTS OF ALLOZYME STUDIES ON EDIBLE MUSHROOMS

The methods for allozyme electrophoresis we have employed are reviewed by May (1992). Other reviews of the methodology for allozyme electrophoresis can be found in Shaw & Prasad (1970), Selander *et al.* (1986), Richardson *et al.* (1986), Pasteur *et al.* (1988), Leary & Booke (1990), Murphy *et al.* (1990), Morizot & Schmidt (1990), Aebersold *et al.* (1987) and Micales *et al.* (1986). We have summarized buffer systems, including electrode buffer, gel buffer and voltage/milliamps requirements, used for multilocus enzyme electrophoresis of enzymes of edible mushrooms in Table 2. In Table 3 we present a list of enzymes, abbreviations, Enzyme Commission number, encoding loci, protein subunit composition, buffer system and mushroom species where allozymes have been studied. Below we discuss each species where allozyme polymorphisms have been examined.

TABLE 2. Buffers used for multilocus enzyme electrophoresis of enzymes of edible mushrooms.

Buffer	Electrode buffer	Gel buffer	Voltage/Milliamps
C[a]	7.687 g citrate (0.04M) 1 liter of water 10.00 ml N-(3-amino-propyl)-morpholine used to adjust to pH 6.1	1 part electrode: 9 parts water	<200 V; 75 mA
K[b]	25 mM Tris	Resolving gel: 375 l mM Tris/HC	30 mA; resolving
	192 mM Glycine pH 8.3	Stacking gel: 125 mM Tris/HCl; pH 6.8	50 mA; stacking
K-2[c]	25mM Tris	Resolving gel: 566 mM Tris/HCl	30 mA; resolving
	125 mM Histidine pH 7.5	Stacking gel: 409 mM Tris/phosphate; pH 5.5	50 mA; stacking

TABLE 2 continued

Buffer	Electrode buffer	Gel buffer	Voltage/Milliamps
KE[d]	Upper electrode 50 mM Tris 383 mM glycine pH 8.3 Lower electrode 63 mM Tris/HCl pH 7.5	Resolving gel: 50 mM Tris/HCl Stacking gel: 50 mM Tris/HCl	200 V 200 V
M[e]	21.81 g tris (0.18M) 6.18 g boric acid (0.1M) 1.17 g EDTA (0.004M) 1 liter of water; 7 adjust to pH 8.	1 part electrode: 3 parts water, pH 8.7	275 V; <75 mA
R[f]	2.51 g lithium hydroxide (0.06M) 18.55 g boric acid (0.3M) 1 liter of water; adjust to pH 8.1	3.63 g tris (0.03M) 0.96 g citrate (0.005M) 10 ml R electrode buffer; pH 8.5	<200 V; <75 mA
S-9[g]	12.10 g tris (0.1M) 11.60 g maleic acid (0.1M) 3.36 g EDTA (0.01M) 2.03 g $MgCl_2$ (0.01M) 6.34 g NaOH (0.12M) 1 liter of water; adjust to pH 8.0	1 part electrode: 19 parts water	85 mA
4[g]	27 g tris (0.223M) 18.07 g citric acid (0.094M) 2.00 g NaOH 1 liter of water; adjust to pH 6.3	0.97 g tris (0.008M) 0.63 g citric acid (0.003M) 0.11 g NaOH 1.00 liter of water; adjust to pH 6.7	<200 V; 75 mA
P[h]	0.3M boric acid 0.05M sodium hydroxide	0.076M tris-(hydroxymethyl)- aminomethane 0.005M citric acid pH 8.65	Not specified

[a]Clayton & Tretiak, 1972; [b]Kulkarni *et al., 1987;* [c]Kulkarni & Kamerath, 1989; [d]Kerrigan & Ross, 1989; [e]Markert & Faulhaber, 1965; [f]Ridgeway *et al., 1970;* [g]Selander *et al., 1971;* [h]Poulik, 1957.

TABLE 3. List of enzymes, abbreviations, Enzyme Commission number, encoding loci, protein subunit composition, buffer system and edible mushroom species where allozymes have been studied.

Enzyme	Abbrev.	EC no.	Loci	Subunits	Buffer	Species
Acid phosphatase	ACP	3.1.3.2	Acp	2	C 4	*V. volvacea* *A. bisporus*
Adenylate kinase	AK	2.7.4.3	Ak	1	M	*L. edodes,* *P. ostreatus*
Alcohol dehydrogenase	ADII	1.1.1.1	Adh	2	K M	*Morchella* spp., *P. ostreatus,* A. bisporus, A. *campestris*
Aldolase	ALD	4.1.2.13	Ald	4	K	*P. ostreatus*
Alkaline phosphatase	AKP	3.1.3.1	Akp	2	K M, P	*P. ostreatus* *Morchella* spp.
Aspartate aminotransferase	AAT	2.6.1.1	Aat	2	C	*A. bisporus,* *A. campestris,* L. edodes, *Pleurotus* spp., *V. volvacea*
Adenosine deaminase	ADA	3.5.4.4	Ada	1	4	*L. edodes,* *V. volvacea*
Catalase	CAT	1.11.1.6	Cat	4	C	*L. edodes*
Diaphorase	DIA	1.6.4.3	Dia	2	C M R K	*L. edodes* *A. campestris* *Pleurotus* spp., V. volvacea *Morchella* spp.
Esterase	EST	3.1.1.1	Est	1	C M R	*V. volvacea* *L. edodes,* *Morchella* spp. *Pleurotus* spp.

TABLE 3 continued

Enzyme	Abbrev.	EC no.	Loci	Subunits	Buffer	Species
Esterase (butylesterase)	ESTB	3.1.1.1	Est-b	1	4	*A. bisporus*
Fumarase	FUM	4.2.1.2	Fum	4	4	*Morchella* spp., *Pleurotus* spp.
					K	*P. ostreatus*
Galactose dehydrogenase	GADH	1.1.1.48	Gadh	2	K	*P. ostreatus*
Glutamic dehydrogenase	GDH	1.4.1.4	Gdh	2	R	*L. edodes*, *Morchella* spp.
Glucokinase	GK	2.7.1.2	Gk	1	C	*A. campestris*
					S-9	*L. edodes*
Glucose-6-phosphate dehydrogenase	G-6-P	1.1.1.49	G6p	2	K	*P. ostreatus*, *Morchella* spp.
					P	*Morchella* spp.
Glucosephosphate isomerase	GPI	5.3.1.9	Gpi	2	4	*A. campestris*, *L. edodes*, *Morchella* spp., *Pleurotus* spp., *V. volvacea*,
					K-2	*Morchella* spp.
					C,M	*Morchella* spp.
Glutathione reductase	GR	1.6.4.2	Gr	2	R	*Pleurotus* spp.
Glutamic pyruvic transaminase	GPT	2.6.1.2	Gpt	2	M	*A. bisporus*, *A. campestris*, *V. volvacea*
β-Glucosidase	B-GLU	3.2.1.21	β-Glu	2	4	*L. edodes*
					KE	*A. bisporus*
					R	*A. campestris*

TABLE 3 continued

Enzyme	Abbrev.	EC no.	Loci	Subunits	Buffer	Species
Glyceraldehyde-3-phosphate dehydrogenase	GAPDH	1.2.1.12	Gapdh	4	C	*Morchella* spp.
					K	*P. ostreatus*
Glutamate oxaloacetate transaminase	GOT	2.6.1.1	Got	2	C	*Morchella* spp.
Guanine deaminase	GDA	3.5.4.3	Gda	2	M	*L. edodes*
Hexose aminadase	HA	3.2.1.30	Ha	1	R	*Pleurotus* spp.
Hexokinase	HK	2.7.1.1	Hk	1	K	*P. ostreatus*
					K-2	*Morchella* spp.
					M, P	*Morchella* spp.
Isocitrate dehydrogenase	IDH	1.1.1.42	Idh	2	4	*Morchella* spp.
					K	*P. ostreatus*
Lactate dehydrogenase	LDH	1.1.1.27	Ldh	4	M	*Morchella* spp.
Leucine aminopeptidase	LAP	3.4.11.1	Lap	1	R	*Morchella* spp. *Pleurotus* spp., *V. volvacea*
					C	*Morchella* spp.
Malate dehydrogenase	MDH	1.1.1.37	Mdh	2	C	*A. campestris*, *Pleurotus* spp., *L. edodes*
Malic enzyme	ME	1.1.1.40	Me	4	C	*Morchella* spp.
Mannitol dehydrogenase	MNDH	1.1.1.138	Mndh	2	K	*P. ostreatus*, *Morchella* spp.
					P	*Morchella* spp.

TABLE 3 continued

Enzyme	Abbrev.	EC no.	Loci	Subunits	Buffer	Species
Mannosephos-phate isomerase	MPI	5.3.1.8	Mpi	1	M	*A. campestris, L. edodes, Morchella* spp., *V. volvacea*
					K	*Morchella* spp.
4-methylum-belliferyl phosphate	MUP	3.1.3.-	Mup	2	R	*Pleurotus* spp.
Nothing dehydrogenase	NDH	1.1.1.-	Ndh	2	4	*Pleurotus* spp.
Nucleoside phosphorylase	NP	2.4.2.1	Np	3	C	*L. edodes, Pleurotus* spp., *V. volvacea*
Peptidase with glycyl-leucine	PEP-GL	3.4.11-13	PepGl	2	R	*A. campestris, L. edodes, Morchella* spp. , *Pleurotus* spp., *V. volvacea*
Peptidase with leucyl-glycyl-glycine	PEP-LGG	3.4.11-13	PepLgg	1	M	*L. edodes, Morchella* spp., *Pleurotus* spp
Peptidase with leucyl-leucyl-leucine	PEP-LLL	3.4.11-13	Pep-1,2	1,2	M	*A. bisporus, A. campestris, L. edodes, V. volvacea*
Peptidase with phenyl-alanyl-proline	PEP-PAP	3.4.11-13	PepPap	2	4,R	*Pleurotus* spp., *V. volvacea*
Phosphoglycerate kinase	PGK	2.7.2.3	Pgk	1	4	*A. campestris, L. edodes, Morchella* spp., *Pleurotus* spp.

TABLE 3 continued

Enzyme	Abbrev.	EC no.	Loci	Subunits	Buffer	Species
Phosphogluconate dehydrogenase	PGD	1.1.1.46	Pgd	2	C	*L. edodes*, *Morchella* spp., *Pleurotus* spp., *V. volvacea*
Phosphogluco- mutase	PGM	2.7.5.1	Pgm	1	S-9	*A. bisporus*, *L. edodes*, *Pleurotus* spp., *V. volvacea*
Inorganic pyrophosphatase	PP	3.6.1.1	Pp	2	S-9	*Pleurotus* spp.
Shikimic dehydrogenase	SKDH	1.1.1.25	Skdh	2	C	*Pleurotus* spp.
Succinate dehydrogenase	SDH	1.2.99.1	Shd	2	K	*P. ostreatus*
Superoxide dismutase	SOD	1.15.1.1	Sod	2	C	*L. edodes*, *Morchella* spp., *V. volvacea*
					K, P	*Morchella* spp.
Triosephosphate isomerase	TPI	5.3.1.1	Tpi	2	4	*Morchella* spp., *Pleurotus* spp.
Xanthine dehydrogenase	XDH	1.1.1.204	Xdh	2	M	*Morchella* spp.

3.1. *Agaricus bisporus*

The nature and extent of genetic diversity within a species is dependent upon its mode of reproduction. In *Agaricus bisporus*, the occurrence of predominantly bispored basidia (heterokaryotic spores) results in a high degree of self-fertility (Lambert, 1929). This mode of reproduction gives rise to reproductively capable single spores that, in the absence of recombination or mutation, are genetically identical to their parental sporocarp.

Population studies on this species were not possible until recently because a suitable means of distinguishing between wild lines and lines that escaped from cultivation was not available (Royse & May, 1982a; 1982b). In addition, minimal resources and effort had been allocated to collecting and preserving wild populations or single individuals collected in the wild. The ability to distinguish a wild population from a population that escaped from cultivation was made possible by the identification and use of allozymes (Royse & May, 1982a; 1982b). Later, with the discovery and identification of wild populations on the west coast of the United States (Kerrigan & Ross, 1989; Kerrigan, 1990) new sources of genetic diversity became available for study. These new sources of germplasm have provided additional insight into the breeding behavior of *A. bisporus*.

The analysis of populations of *A. bisporus* from California collections suggests that genetic diversity in this species has been fostered by sexual outcrossing (Kerrigan, 1990). If additional lines of evidence are shown to support this suggestion, then a much more diverse source of germplasm may still be available in other parts of North America and the world. Kerrigan (1990) has emphasized the importance of the recovery and analysis of more samples. A more thorough quantification of genetic diversity would be possible using more markers. Such markers may include additional allozymes (Royse & May, 1989; Kerrigan & Ross, 1989), restriction fragment length polymorphisms (Summerbell *et al.*, 1989; Loftus *et al.*, 1988, Castle *et al.*, 1987; Hintz *et al.*, 1985), random amplified polymorphic DNA, or molecular sequence markers. These markers could also add to the genetic map of linkage groups already started for this species by Spear *et al.* (1983) as shown in Fig. 1.

3.2. *Agaricus campestris*

A single natural population of *A. campestris* was examined by allozyme electrophoresis. May & Royse (1982a) found an average heterozygosity of 0.281 for this species which is at the high end for organisms examined electrophoretically (Powell, 1975; Nevo, 1978). Royse & May (1982a) suggested that this species may have a large population size based on the relatively high heterozygosity values observed. Moreover, they suggested that the lack of fit to Hardy-Weinberg expectations for three loci (Dia, Mpi, and Pep-LLL-2) is probably due to the large number of alleles observed at each of the above loci (4, 11, and 6, respectively) and to the relatively small sample sizes examined (63, 64, and 65, respectively). Since this species is known to form fairy rings, it is also possible that the persistence of a common dikaryon in the soil for many years may have led to the departure from random mating expectations.

Comparisons of numbers of polymorphic loci found in *A. campestris* with the numbers of polymorphisms in *A. bisporus* and other edible fungi are presented in Table 4. The proportion of polymorphic loci is higher (0.87) for *A. campestris* than for all other species examined.

TABLE 4. Number of loci examined, number of polymorphic loci and proportion of polymorphic loci derived from allozyme studies of various species of edible fungi.

Species	No. of Loci Examined	No. of Loci Polymorphic	Proportion of Polymorphic Loci {P}	References
Agaricus bisporus	33	12	0.36	Royse & May (1982a; 1989); Kerrigan (1990)
A. campestris	15	13	0.87	May & Royse (1982a)
Lentinula edodes	31	23	0.74	Royse *et al.* (1983a,b); Royse & May (1987); Bowden *et al.* (1991a)
Morchella spp.	20	16	0.80	Royse & May (1990); Yoon *et al.* (1990)
Pleurotus spp.	47	36	0.77	Kulkarni *et al.* (1987); May & Royse (1988); May *et al.* (1988)
Volvariella volvacea	17	5	0.29	Royse *et al.* (1987)

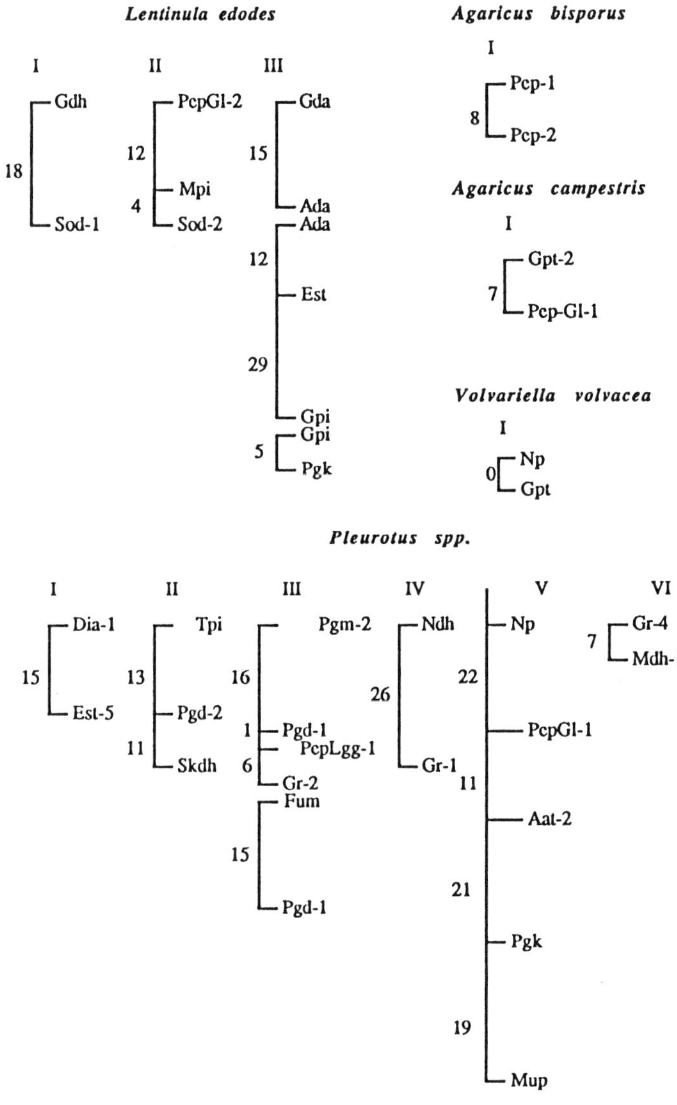

FIGURE 1. Genetic maps of allozyme linkage groups in one genus and four species of edible mushroom. Map distances are in centimorgans (unadjusted recombination percentages).

3.3. *Lentinula edodes*

Allozyme studies on natural populations of *L. edodes* have not been reported. Analysis of available lines have included some wild isolates of *L. edodes* but these have been examined in conjunction with commercial cultivars. Nevertheless, the proportion of polymorphic loci is relatively high. Out of 31 loci examined in various studies (Royse *et al.,* 1983a, 1983b, Royse & May, 1987; Bowden *et al.,* 1991a) 23 loci have been reported to be polymorphic ({P}=0.74). Such a high proportion of polymorphic loci would indicate that there is considerable genetic diversity in this species.

Chi-square tests for single locus segregation of *L. edodes* in two allozyme studies (Royse *et al.,* 1983a; Bowden *et al.,* 1991a) has revealed that only Aat-1 had a significant departure from random segregation. Royse *et al.* (1983a) postulated that a locus tightly linked to the Aat-1 locus could carry a lethal allele. They suggested that only recombinant spores without the lethal allele would be viable. Using recombination frequencies, Royse *et al.* (1983a) mapped the locus carrying the lethal allele and Aat-1 at 4 centimorgans. In an examination of more progeny in another cross using the same WC131 parent, Bowden *et al.* (1991) mapped the lethal allele and the Aat-1 locus at 17 centimorgans. Additional work is needed to confirm this hypothesis because there appears to be no general association between low germination rate and skewed allele ratios (Bowden *et al.,* 1991b).

Royse & May (1987) have cataloged lines of *L. edodes* according to allozyme data from 11 loci. To date, 38 genotypic classes have been recognized. However, with additional allozyme markers now available, the number of genotypes available in the Pennsylvania State University Mushroom Culture Collection (PSUMCC) may now increase. Furthermore, the successful hybridization of two highly divergent lines of *L. edodes* has allowed the production of a genetic linkage map (Fig. 1) of eight allozyme encoding loci (Bowden *et al.,* 1991a). This map will aid in the monitoring of both monogenic and quantitatively inherited traits in *L. edodes.*

3.4. *Morchella* spp.

Interest in the cultivation of morels has increased as a result of recent biochemical and genetic studies (Ower, 1982; Ower *et al.,* 1986; 1988; Gessner *et al.,* 1987; Kulkarni & Kamerath, 1989; Volk & Leonard, 1989;

Yoon *et al.*, 1990; Royse & May, 1990). The genetic life history of the genus still has not been clearly defined. Allozyme data seem to indicate that many ascocarps may be haploid. For example, Yoon *et al.* (1990) examined allozyme activity in field-collected ascocarp tissue, stipe cultures from ascocarps and single-ascospore-derived mycelial cultures. All stipe cultures and ascospore cultures examined were apparently haploid (i.e., no heterozygous loci were observed). Royse & May (1990) tested 31 enzymes for electrophoretic resolution and variability in six species of *Morchella.* Twelve enzymes were sufficiently resolved to allow their use for species comparisons. No heterozygous loci were reported in their study. If ascocarp tissue is diploid, the observation of some heterozygous loci would be expected in some of the ascocarp tissue cultures and field-collected samples. Data from Volk & Leonard (1989) show that heterokaryon formation occurs between monoascosporous stains of *Morchella.* It is not clear, therefore, why heterozygous individuals were not observed in the allozyme studies. Additional allozyme investigations are warranted on well-defined crosses.

The status of *Morchella* taxonomy makes the application of cultivation and breeding methodologies more difficult. It is important to identify species correctly because optimal cultural husbandry probably will differ for each. Effective genetic modifications will require knowledge of the systematic relationships of isolates used in matings. Royse & May (1990) have shown that a number of species may easily be misidentified. Because allozyme analysis allows groupings based on genetic similarities, more rigorous morphological scrutiny may be directed toward those isolates that are more highly divergent. It is possible that current morphological characteristics used to identify field isolates of *Morchella* spp. are not adequate and should be reexamined.

Gessner *et al.* (1987) have proposed the possibility of separate gene pools or distinct species for one population identified morphologically as *M. deliciosa* and one population identified morphologically as *M. esculenta* collected only a few hundred meters from each other in Plymouth, Illinois. Nei's Unbiased Identity Values calculated for these two populations in Illinois was 0.89. The data from Royse & May (1990) support their suggestion that the two populations are merely part of a highly variable species-complex, as judged by allozymic data and thus not worthy of specific recognition.

3.5. *Pleurotus* spp.

The fungal genus *Pleurotus* contains a number of edible species that can be grown on a large variety of agricultural by-products. Worldwide interest in these mushrooms continues to grow because of increased consumer demand, the relative ease of cultivation and the ability to cultivate them on wide variety of substrates (Chang & Hayes, 1978; Chang & Quimio, 1982; Royse & Schisler, 1987; Quimio *et al.*, 1990).

In general, *Pleurotus* isolates of different species do not readily cross. However, May *et al.* (1988) reported that isolates identified morphologically as *P. ostreatus* (WC518) and *P. sapidus* (WC528) were able to hybridize with an isolate of *P. sajor-caju* (WC537). Allozyme variability between these isolates supported their taxonomic status. *Pleurotus sajor-caju* (WC537) originally was isolated from the foot hills of the Himalayas (India) while *P. sapidus* (WC528) and *P. ostreatus* (WC518) were wild collections from British Columbia. These crosses are unusual and satisfying in that they offer a hybridizing mechanism other than protoplast fusion (Yoo *et al.*, 1984; Toyomasu *et al.*, 1986) to produce unique genetic combinations between species (May *et al.*, 1988). May *et al.* (1988) suggested that the mating incompatibilities of these species may break down after sufficient evolutionary (geographic) separation. Their results (May *et al.*, 1988) suggest that there may be other *Pleurotus* species combinations that will be compatible.

Nine of the enzymes in the May *et al.* (1988) study were coded by more than a single locus. Four of the seven enzymes (phosphoglucomutase [PGM], glutathione reductase [GR], peptidase with leucyl-glycyl-glycine [PEP-LGG], and phosphoglucanate dehydrogenase [PGD]) were polymorphic and were examined for nonrandom association between the duplicate loci (e.g., *Pgm-1* with *Pgm-2*) and between the duplicated enzyme systems (e.g. *Pgm-1* and *Pgm-2* with *Pgd-1* and *Pgd-2*). May *et al.* (1988) found that none of the four duplicated enzyme systems were linked to each other, and suggested that the duplicated systems probably did not arise by recent tandem duplication events. These researchers also found that the four duplicated systems were not part of duplicated linkage groups. Such duplicated linkage groups would be indicative that they could have arisen by chromosomal or genomic duplication. The results reported by May *et al.* (1988) suggest that duplications are either ancient or that chromosomal rearrangements are rapid within these fungi.

Slezec (1984) reported chromosome numbers of $n = 12$, 13, and 14 in three lines of *P. eryngii*. Based on relative DNA content data, Bresinsky *et al.* (1987) estimated chromosome numbers for an additional seven species of *Pleurotus* to range from $n = 6$ to $n = 18$ (most being 8-10). The six linkage groups and six unlinked loci, identified by May *et al.* (1988), should mark most of the chromosomes in the *Pleurotus* genome. A linkage map of this genus is presented in Fig. 1. Comparative gene mapping should be possible between the generalized *Pleurotus* genome and *L. edodes* once we add other DNA specific markers.

3.6. *Volvariella volvacea*

The paddy straw mushroom, *V. volvacea*, is the third most important commercially cultivated mushroom following *A. bisporus* and *L. edodes*, accounting for about 8% of the total world production of edible mushrooms. Mainland China and Taiwan together account for about 90% of the world-wide production of this species (Chang, 1987).

Improvements in yield of *V. volvacea* through selective breeding have been impeded by a lack of basic genetic information needed to develop rational breeding programs. Until recently, it was unclear as to whether basidiospores were haploid or heterokaryotic. Since approximately 50% of the single basidiospore isolates are fertile, this criterion could not be used to help establish the ploidy of the single spores. A true sexual process in *V. volvacea* was difficult to verify without the presence of codominantly expressed genetic markers such as allozymes. Royse *et al.* (1987) examined 19 lines of *V. volvacea* gathered from world-wide sources for allozyme activity. Twelve loci were found to be monomorphic while five loci were found to be polymorphic. Among the 19 lines examined, nine unique genotypes were distinguishable based on multilocus comparisons.

Through the use of a well-defined cross in this species Royse *et al. (1987)* were able to examine the genetic behavior of this species. Based on data from single and joint segregation studies of allozyme loci, Royse *et al.* (1987) were able to confirm that basidiospores of this species are haploid and that homokaryons may mate to yield heterokaryons. Population studies are needed with this species to determine the extent of genetic variation in wild populations. Apparently this species is widespread in the tropics allowing the possibility of examining several geographically separated populations.

4. FUTURE OUTLOOK

We suggest that multilocus enzyme electrophoresis will continue to play a very significant role in genetic studies of edible mushrooms for some time to come. We suggest the continued value of their role even in the face of the growing emphasis on direct examinations of DNA variation (e.g., RFLP's, RAPD, and DNA sequencing). In fact, the greatest power will undoubtedly come from combining allozyme and DNA data. The following areas of application of allozyme data should be carefully examined:

4.1. Species Identification

One of the things we have learned through the allozyme studies is the level of misidentification of species in some genera (e.g., *Pleurotus* and *Morchella*). Extensive surveys of allozyme variation should be conducted in highly speciose genera to determine species specific markers and to delineate and systematically group the member species.

4.2. Germplasm Collections

Germplasm collections need to be evaluated for extant variation and to determine which lines are truly different from one another. The historical practice of sharing isolates among laboratories results in considerable duplication of lines in collections.

4.3. Natural populations

One of the more exciting areas of research will be assessing genetic variation in those species with natural populations. These populations harbor the raw genetic variation available for selection programs.

4.4. Linkage studies

As RFLP and RAPD linkage maps are developed for mushroom species, we need a mechanism for comparing maps. Allozymes provide such a mechanism. The locus coding for the enzyme glucosephosphate isomerase (GPI) is the same in all fungi examined.

4.5. Confirmation of crosses

Allozymes permit one of the easiest, cheapest methods of confirming when an actual cross is made between two single spore derived lines. This method already is being used commercially at some spawn makers in the United States.

4.6. Patent labeling

Labeling of new lines of edible mushrooms demands unambiguous genetic markers that are unique to the new line. The marker must be highly variable and codominant. Allozymes, RAPD, DNA sequences, and variable number tandem repeat (VNTR) loci are all valid candidates for this role.

REFERENCES

AEBERSOLD, P.B., WINANS, G.A., TEEL, D.J., MILNER, G.B. & UTTER, F.M. (1987). *Manual for Starch Gel Electrophoresis: A Method for Detection of Genetic Variation.* National Oceanic and Atmospheric Administration Technical Report NMFS 61. Washington, D.C.: U. S. Dept. of Commerce

BRESINSKY, A., FISCHER, M., MEIXNER, B. & PAULUS, W. (1987). Speciation in *Pleurotus*. *Mycologia* **79**, 234-245.

BOWDEN, C.G., ROYSE, D.J. & MAY, B. (1991a). Linkage relationships of allozyme encoding loci in *Lentinula edodes*. *Genome* **34** 652-657.

BOWDEN, C.G., ROYSE, D.J. & MAY, B. (1991b). Linkage relationships of one mating factor locus with thirteen allozyme encoding loci in shiitake, *Lentinula edodes*. *Mycological Society of America Newsletter* **42**: 7.

CASTLE, A.J., HORGEN, P.A. & ANDERSON, J.B. (1987). Restriction fragment length polymorphisms in the mushrooms *Agaricus brunnescens* and *Agaricus bitorquis*. *Applied and Environmental Microbiology* **53**, 816-822.

CHANG, S.T. (1987). World production of cultivated edible mushrooms in 1986. *Mushroom Journal of the Tropics* **7**, 117-120.

CHANG, S.T. & HAYES, W.A. (1978). *The Biology and Cultivation of*

Edible Mushrooms. New York.: Academic Press.

CHANG, S T. & QUIMIO, T.H. (1982). *Tropical Mushrooms: Biological Nature and Cultivation Methods.* Hong Kong: The Chinese University Press.

CLAYTON, J.W. & TRETIAK, D.N. (1972). Amine-citrate buffers for pH control in starch gel electrophoresis. *Journal of the Fish Research Board of Canada* **29**, 1169-1172.

GESSNER, R.V., RAMANO, M.A. & SCHULTZ, R.W. (1987). Allelic variation and segregation in *Morchella deliciosa* and *M. esculenta. Mycologia* **79**, 683-687.

HARRIS, H. (1966). Enzyme polymorphism in man. *Proceedings of the Royal Society of London, Series B* **164**, 298-310.

HINTZ, W.E., MOHAN, M., ANDERSON, J.B. & HORGEN, P.A. (1985). The mitochondrial DNAs of *Agaricus:* heterogeneity in *A. bitorquis* and homogeneity in *A. brunnescens. Current Genetics* **9**, 127-132.

KERRIGAN, R.W. (1990). Evidence of genetic divergence in two populations of *Agaricus bisporus. Mycological Research* **94**, 721-733.

KERRIGAN, R.W. & ROSS, I.K. (1989). Allozymes of a wild *Agaricus bisporus* population: new alleles, new genotypes. *Mycologia* **81**, 433-443.

KULKARNI, R. & KAMERATH, C.D. (1989). Isozyme analysis of *Morchella* species. *Mushroom Science* **12**, 451-457.

KULKARNI, R., KAMERATH, C.D. & ALLRED, K.L. (1987). Genetic diversity between isolates of *Pleurotus ostreatus* as revealed by isozyme analysis. In *Cultivating Edible Fungi,* pp 178-182. Edited by P.J. Wuest, R.B. Beelman, and D.J. Royse. New York: Elsevier.

LAMBERT, E.B. (1929). The production of normal sporophores in monosporus cultures of *Agaricus bisporus. Mycologia* **21**, 333-335.

LEARY, R.F. & BOOKE, H.E. (1990). In *Methods for Fish Biology* pp 141-170. Bethesda, MD: American Fisheries Society.

LEWONTIN, R.C. & HUBBY, J.L. (1966). A molecular approach to the study of genic heterozygosity in natural populations. II. Amount of variation and degree of heterozygosity in natural populations of *Drosophila pseudoobscura. Genetics* **54**, 595-609.

LOFTUS, M.G., MOORE, D. & ELLIOTT, T.J. (1988). DNA polymorphisms in commercial and wild strains of the cultivated mushroom, *Agaricus bisporus. Theoretical and Applied Genetics* **76** 712-718.

MARKERT, C.L. & FAULHABER, I. (1965). Lactate dehydrogenase isozyme patterns of fish. *Journal of Experimental Zoology* **159**, 319-332.

MAY, B. (1992). Starch gel electrophoresis of allozymes. In *Molecular Genetic Analysis of Populations: A Practical Approach*, Edited by A. R. Hoelzel. Oxford: IRL Press. In press.

MAY, B. & ROYSE, D.J. (1981). Applications of the electrophoretic methodology to the elucidation of genetic life histories of edible mushrooms. *Mushroom Science* **11**, 799-817.

MAY, B. & ROYSE, D.J. (1982a). Genetic variation and joint segregation of biochemical loci in the common meadow mushroom, *Agaricus campestris*. *Biochemical Genetics* **20**, 1165-1173.

MAY, B. & ROYSE, D. (1982b). Confirmation of crosses between lines of *Agaricus brunnescens* by isozyme analysis. *Experimental Mycology* **6**, 283-292.

MAY, B. & ROYSE, D. (1988). Interspecific allozyme variation within the fungal genus *Pleurotus*. *Transaction of the British Mycological Society* **90**, 29-36.

MAY, B., HENLEY, K.J., FISHER, C.G. & ROYSE, D.J. (1988). Linkage relationships of 19 allozyme encoding loci within the commercial mushroom genus *Pleurotus*. *Genome* **30**, 888-895.

MICALES, J.A., BONDE, M.R. & PETERSON, G.L. (1986). The use of isozyme analysis in fungal taxonomy and genetics. *Mycotaxon* **27**, 405-449.

MORIZOT, D.C. & SCHMIDT, M.E. (1990). In *Electrophoretic and Isoelectric Focusing Techniques in Fisheries Management*, pp 23-80. Boca Raton, FL: CRC Press.

MOUCHES, C., DUTHIL, P., BOVE, J.M., VALJALO, J. & DELMAS, J. (1979). Characterisation des champignons superieurs par electrophorese de leurs proteins. *Mushroom Science* **10**, 491-503.

MURPHY, R.W., SITES, J.W., BUTH, D.G. & HAUFLER, C.H. (1990). *Molecular Systematics*, pp 45-126. Edited by D.M. Hillis and C. Moritz. Sunderland, MA. Sinauer Associates.

NEVO, E. (1978). Genetic variation in natural populations: patterns and theory. *Theoretical and Population Biology* **13**, 121-145.

OWER, R. (1982). Notes on the development of the morel ascocarp: *Morchella esculenta*. *Mycologia* **74**, 142-144.

OWER, R.D., MILLS, G.L. & MALACHOWSKI, J.A. (1986). Cultivation of *Morchella.* U.S. Patent No. 4,594,809.

OWER, R.D., MILLS, G.L. & MALACHOWSKI, J.A. (1988). Cultivation of *Morchella.* U.S. Patent No. 4,757,640.

PARANJPE, M.S., CHEN, P.K. & JONG, S.C. (1979). Morphogenesis of *Agaricus bisporus*: changes in proteins and enzyme activity. *Mycologia* **71**, 469-478.

PASTEUR, N., PASTEUR, G., BONHOMME, F., CATALAN, J. & BRITTON-DAVIDIAN, J. (1988). *Practical Isozyme Genetics.* Chichester: Ellis Horwood.

POULIK, M.D. (1957). Starch gel electrophoresis in a discontinuous system of buffers. *Naturc* **180**, 1477-1479.

POWELL, J.R. (1975). Protein variation in natural populations of animals. In *Evolutionary Biology*, pp. 79-119. Edited by T. Dobzyhansky., M.K. Hecht, and W.C. Steere. New York: Plenum.

QUIMIO, T.H., CHANG, S.T. & ROYSE, D.J. (1990). *Technical Guidelines for Mushroom Production in the Tropics.* FAO Plant Production and Protection Paper 106.

RAPER, C.A. & KAYE, G. (1978). Sexual and other relationships in the genus *Agaricus. Journal of General Microbiol*ogy **105**, 135-151.

RICHARDSON, B.J., BAVERSTOCK, P.R. & ADAMS, M. (1986). *Allozyme Electrophoresis: a Handbook for Animal Systematics and Population Studies.* Sydney: Academic Press.

RIDGEWAY, G.J., SHERBURNE, S.W. & LEWIS, R.D. (1970). Polymorphism in the esterases of Atlantic herring. *Transaction of the American Fish Society* **99**, 147-151.

ROYSE, D.J. & MAY, B. (1982a). The use of isozyme variation to identify genotypic classes of *Agaricus brunnescens. Mycologia* **74**, 93-102.

ROYSE, D.J. & MAY, B. (1982b). Genetic relatedness and its application in selective breeding of *Agaricus brunnescens. Mycologia* **74**, 569-575.

ROYSE, D.J. & MAY, B. (1987). Identification of shiitake genotypes by multilocus enzyme electrophoresis: catalog of lines. *Biochemical Genetics* **25**, 705-716.

ROYSE, D.J. & MAY, B. (1989). Identification and use of three new biochemical markers in *Agaricus bisporus. Agricultural and Biological Chem*istry **53**, 2861-2866.

ROYSE, D.J. & MAY, B. (1990). Interspecific allozyme variation among *Morchella* spp. and its inferences for systematics within the genus.

Biochemical Systematic Ecology **18**, 475-479.

ROYSE, D.J. & SCHISLER, L.C. (1987). Yield and size of *Pleurotus ostreatus* and *Pleurotus sajor-caju* as effected by delayed-release nutrient. *Applied Microbiology and Biotechnology* **26**, 191-194.

ROYSE, D.J., SPEAR, M.C. & MAY, B. (1983a). Single and joint segregation of marker loci in the shiitake mushroom, *Lentinus edodes*. *Journal of General and Applied Microbiology* **29**, 217-222.

ROYSE, D.J., SPEAR, M.C. & MAY, B. (1983b). Cell line authentication and genetic relatedness of lines of the shiitake mushroom, *Lentinus edodes*. *Journal of General and Applied Microbiology* **29**, 205-216 .

ROYSE, D.J., JODON, M.H., ANTOUN, G.G. & MAY, B. P. (1987). Confirmation of intraspecific crossing and single and joint segregation of biochemical loci of *Volvariella volvacea*. *Experimental Mycology* **11**, 11-18.

SELANDER, R.K., SMITH, M.H., YANG, S.Y., JOHNSON, W.E. & GENTRY, J.B. (1971). Biochemical polymorphism and systematics in the genus *Peromyscus* I. Variation in the old-field mouse (*Peromyscus polionotus*). *Studies in Genetics* **6**, 49-90.

SELANDER, R.K., CAUGANT, D.A., OCHMAN, H., MUSSER, J.M., GILMOUR, M.N. & WHITTAM, T.S. (1986). Methods of multilocus enzyme electrophoresis for bacterial populations genetics and systematics. *Applied and Environmental Microbiology* **51**, 873-884.

SHAW, C.R. & PRASAD, R. (1970). Starch gel electrophoresis of isozymes - a compilation of recipes. *Biochemical Genetics* **4**, 297.

SHECHTER, Y., HALL, R., OELSHLEGAL, JR., F.S., STAHMAN, R.D., SNIDER, M.A., FRANKE, R.G., HOLMAN, R.W., REDDY, M.M. & GARBER, E.D. (1973). Symposium on the use of electrophoresis in the taxonomy of algae and fungi. *Bulletin of the Torrey Botanical Club* **100**, 253-312.

SLEZEC, A.-M. (1984). Variabilite du nombre chromosomique chez les pleurotes des ombelliferes. *Canadian Journal Botany* **62**, 2610-2617.

SPEAR, M.C., ROYSE, D.J. & MAY, B. (1983). Atypical meiosis and joint segregation of biochemical loci in *Agaricus brunnescens*. *Journal of Heredity* **74**, 417-420.

SPEITH, P.T. (1975). Population genetics of allozyme variation in *Neurospora intermedia*. *Genetics* **80**, 785-805.

SUMMERBELL, R.C., CASTLE, A.J., HORGEN, P.A. & ANDERSON, J.

B. (1989). Inheritance of restriction fragment length polymorphisms in *Agaricus brunnescens*. *Genetics*, **123**, 293-300.

TOYOMASU, T., MATSUMOTO, T. & MORI, K.-I. (1986). Interspecific protoplast fusion between *Pleurotus ostreatus* and *Pleurotus salmoneastramineus*. *Agricultural and Biological Chemistry* **50**, 223-225.

VOLK, T.J. & LEONARD, T.J. (1989). Experimental studies on the morel. I. Heterokaryon formation between monoascosporous strains of *Morchella*. *Mycologia* **81**,523-531.

YOO, Y.-B., BYUN, M.-O., GO, S.-J., YOU, C.-H., PARK, Y.H. & PEBERDY, J.F. (1984). Characteristics of fusion products between *Pleurotus ostreatus* and *Pleurotus florida* following interspecific protoplast fusion. *Korean Journal of Mycology* **12**, 164-169.

YOON, C-S., GESSNER, R.V. & RAMONO, M.A. (1990). Population genetics and systematics of the *Morchella esculenta* complex. *Mycologia* **82**, 227-235.

CHAPTER 12

METHODS OF GENETIC MANIPULATION IN *COPRINUS CINEREUS*

Patricia J. Pukkila

Department of Biology, University of North Carolina,
Chapel Hill, 27599, U.S.A.

1. INTRODUCTION

Many species of basidiomycetous fungi are of considerable economic importance. This group of fungi includes edible mushrooms, wood rotters, and serious crop pathogens. However, the application of molecular genetic techniques to improvement or control of these fungi has lagged behind progress in other areas due to the lack of a suitable model system that can be manipulated in the laboratory. The agaric *Coprinus cinereus* offers many technical advantages and appears to be particularly relevant to ongoing studies that use hymenomycetes of greater commercial value.

There are several aspects to the life history of *C. cinereus* that make it an attractive system with direct relevance to edible species. First, development is relatively rapid in *C. cinereus*. The life cycle from basidiospore germination, mating, fruit body formation, and production of the next generation of basidiospores takes only two weeks, and all stages can be cultured in the laboratory using simple media and equipment (Moore & Pukkila, 1985). In comparison to edible fungi, *C. cinereus* development is more rapid, more predictable, and more easily manipulated by varying simple light and nutritional cues. However, the underlying morphogenetic processes appear to be identical in these basidiomycetes, so genetic information concerning metabolic and developmental pathways in *C. cinereus* is likely to be directly relevant to these other species.

To date, most studies using *C. cinereus* have concentrated on four issues of particular interest: control of mating behavior, mitochondrial gene transmission and recombination, fruit body formation, and meiotic chromosome behavior (reviewed in Pukkila & Casselton, 1991). This chapter will concentrate on genetic techniques that have been developed to facilitate those studies, since many of these methods should be suitable for other species as well. I will first briefly describe conventional mutant isolation and detail a simple storage method that is suitable when large numbers of strains are utilized. I will then turn to the application of simple mapping procedures, both genetic and physical, and illustrate how these can be used to eliminate undesirable characteristics in particular strains. Finally, I will describe molecular manipulations that are now possible following the development of a DNA-mediated transformation system.

2. MUTANT ISOLATION

2.1. Selection and Screening Using Haploid Strains

Mutant isolation in *C. cinereus* is greatly facilitated by the fact that the monokaryotic mycelium produces copious numbers of asexual spores (oidia). These are seldom airborne, but instead collect in drops of fluid and can be recovered by scraping the mycelial surface. Particularly large numbers of oidia are produced by most strains when grown on yeast-malt-glucose (YMG) medium (Rao & Niederpruem, 1969). Thus, filtration enrichment, first developed for mutant isolation in *Ophiostoma multiannulatum* (Fries, 1947) is quite suitable in *C. cinereus*. Ultraviolet light and *N*-methyl-*N'*-*N*-nitrosoguanidine are the mutagens that have been used most widely. The mutagenized oidia are allowed to germinate in supplemented medium lacking the component of interest. Prototrophic oidia grow, while those with genetic defects that prevent growth in the absence of the missing component fail to germinate. By filtering the suspension at regular intervals, a considerable enrichment for mutants in the desired pathway is achieved. These are rescued by plating on fully supplemented media.

Such methods, coupled with the use of particular metabolic inhibitors including fluoroacetate (Casselton & Casselton, 1974), cyclohexamide (North, 1982), 5-fluoroindole (Veal & Casselton, 1985), and benomyl (Kamada *et al.*, 1989*a*) have allowed characterization of mutations in over 100 different genes. Recently, benomyl resistant variants were used to identify the structural genes for a major β–tubulin (*benA*) and an α–tubulin (*benC*) (Kamada *et al.*, 1989*b*; 1990). Mutant alleles were recovered that block the ability of homokaryons to receive nuclei during mating, although they could donate nuclei to a compatible homokaryon. A high rate of dedikaryotization was noted, which suggests that the genes may play an important role in nuclear behavior during conjugate nuclear division of the dikaryon. Surprisingly, nuclear migration into developing basidiospores, and nuclear migration during basidiospore germination appeared to be normal, suggesting that alternate pathways control those movements (Kamada *et al.*, 1989b).

Temperature sensitive mutations have also been sought in *C. cinereus*. Among the most interesting of these were obtained in a collection of mutants recovered by Kamada and co-workers that confer defects in hyphal tip growth at elevated temperature, but allow normal growth at the permissive temperature (Kamada *et al.*, 1984). These should prove most valuable in understanding the molecular controls of hyphal tip morphogenesis.

2.2. Screening Using *AmutBmut* Strains

Fruit body morphogenesis has attracted numerous physiological studies, but genetic analysis was hampered by the limitations imposed by the mating system. Some progress was made using dominant developmental mutants (Takemaru & Kamada, 1972; Gibbins & Lu, 1982), but by far the most useful screens have used the *Amut Bmut* strains of Swamy *et al.* (1984). Single spore isolates of such strains will form homokaryotic fruiting bodies, so mutagenized haploid cells can be screened for those that harbor developmental mutants without the necessity to first mate the haploid to a compatible strain. The extensive labor that would be involved in mutagenizing, outcrossing, and backcrossing (with only one in four matings being productive due to the tetrapolar mating system) to uncover recessive mutants is thus avoided. Of considerable importance is the fact that *AmutBmut* strains will donate nuclei to monokaryons that contain conventional mating types. After carrying out such di-mon matings, one can detect strains that carry the recessive developmental mutations and conventional mating types using di-mon backcrosses to the original *Amut Bmut* mutant isolate. Thus, the new mutations can be recovered in conventional genetic backgrounds even when the developmental defect prevents the production of viable haploid spores. As long as the mutations are recessive, extensive genetic analysis can be performed.

Several developmental pathways have been revealed using these methods. Fruit body morphogenesis can be disrupted by single genes prior to hyphal aggregation into *knots* (*knt*$^-$), *prim*ordium formation (*prm*$^-$), primordium *mat*uration (*mat*$^-$), or basidio*spo*re formation (*spo*$^-$) (Kanda & Ishikawa, 1986; A. B. Maynard & P.J. Pukkila, unpublished observations). Separate pathways appear to control cap *exp*ansion and autolysis (*exp*$^-$) and stipe *el*o*n*gation (*eln*$^-$) (Takemaru & Kamada, 1972).

The genetic pathways that control the synchronous meiotic process and subsequent basidiospore formation have been dissected in some detail. Single mutations causing arrest at several discrete stages have been isolated (Zolan *et al.*, 1988; Kanda *et al.*, 1989). The *rad*iation-sensitive mutation *rad3-1* was of considerable interest (Zolan *et al.*, 1988). Mutant strains failed to condense and pair their chromosomes properly, although surviving spores exhibited normal amounts of recombination. In another study (Kanda *et al.*, 1990), it was concluded that chromosome pairing and duplication of the spindle pole body can occur in the absence of premeiotic DNA replication. It will be of interest to determine the relationships between genes such as these whose failure to function normally leads to a complete failure of spore formation, and another set of basidiospore development (*bad*$^-$) mutants in which spores are formed, but are not viable. The mechanisms that underlie coordination of the meiotic pathway and spore formation pathway could be analyzed in this fashion. It should also be possible to use DNA mediated transformation to clone many of these genes for structure-function studies. Finally, homologous genes are undoubtedly present in edible species, and their analysis might facilitate certain types of strain modification.

2.3. Storage of *C. cinereus* Strains

Many strains are generated in mutant hunts, and a fast and reliable storage method is needed to avoid loss of interesting variants. We have found that strains can be stored for several years in 15% glycerol at -70°C. We use 1.2 ml NUNC cryotubes (Vanguard International, Inc.) and place 0.75 ml of 15% glycerol (w:w in distilled water) in each. The strain to be stored is grown on an YMG agar plate, and a tungsten needle is used to cut blocks of mycelium plus the underlying agar into small squares ($0.5cm^2$). A few chunks are placed in each vial. The agar is not necessary, but it facilitates recovery of the piece after thawing. Vials can be frozen and thawed several times with no loss of viability. In our hands, this method is more simple and much faster than silica gel storage.

3. RELATING GENETIC AND PHYSICAL MAPS

A combination of random spore and tetrad analysis has been used to develop two linkage maps for *C. cinereus*. Since these were developed largely due to experiments carried out in Japan or in Britain, I will refer to the maps accordingly. In the Japanese map (Takemaru, 1982), 38 markers were mapped relative to each other and used to define 7 linkage groups. Centromere position was not reported. In the British map (North, 1987), over 100 markers were used to define 10 linkage groups, and 9 centromeres were distinguished. However, not all the markers were mapped relative to their nearest neighbors. Some of these data derive from reports in which *C. cinereus* was called *C. lagopus* or *C. macrorhizus f. microsporus* Hongo, but these (and also *C. fimentarius* are now called *C. cinereus* (Shaeff, ex Fr.) S.F. Gray.

Several technical issues have impeded progress in assembling a more complete map. First, until recently, little attempt was made to fuse the separate maps by determining which markers are in fact allelic or linked. Second, the use of non-isogenic strains can lead to reduced spore viability, and so tetrad analysis can be laborious. Finally, strains with multiple auxotrophic markers tend to have difficulty fruiting. These difficulties are not insurmountable. We have begun to fuse the two maps by incorporating markers from both maps in complementation tests and crosses, and the current status of this project is summarized in section 3.1. below. To overcome the problems with spore viability, we have adopted a standard strain, and recovered compatible strains with an identical karyotype that are near-isogenic with the standard. Tetrad analysis with such strains can quickly reveal deleterious mutations in the strain background that interfere with fruiting (described in section 3.2. below). Finally, to circumvent the problems associated with the use of multiple auxotrophic markers, we have turned to a variety of molecular methods that allow physical maps to be constructed (described in section 3.3.).

3.1. A Unified Genetic Map

Table 1 lists our current suggestions for a unified nomenclature for *C. cinereus* linkage groups. We have chosen strain 218 as our standard strain (Binninger *et al.*, 1987) because DNAs from each of its chromosomes can be resolved easily using electrophoretic methods that are described under section 3.3. below. We have numbered the chromosomes of this strain in order of decreasing size. Both cytological and molecular methods have been used to estimate the length of each chromosome in megabases (mb). Relative chromosome sizes were estimated in several strains by both electron microscopy (Holm *et al.*, 1981) and light microscopy (Pukkila & Lu, 1985). Since the total genome size is 37.5 mb (Dutta, 1974), the approximate DNA content of each chromosome could be calculated. We have also estimated the relative lengths of the chromosomal DNA in several strains using electrophoretic methods (described in section 3.3.). The consensus from all these estimates is recorded in Table 1.

Work in progress should soon allow relationships to be established between Japanese groups II, V and VI (using *B*, *ARG1* and *HIS2*) and British groups II, V and IV (using *B*, *ADE6* and *MET9*). The cloned probes used to establish the indicated physical linkages are described in Pukkila & Casselton (1991) with the exception of the RFLP marker on chromosome XIII (O7XIII-1E4, M. Zolan, personal communication). Cloned probes for the remaining bands should facilitate assignment of the smaller linkage groups to chromosomes.

3.2. Tetrad Analysis with Near-Isogenic Strains

Tetrad analysis offers several advantages over random spore analysis: centromere linkage can be detected easily, only a small number of progeny need to be analyzed to demonstrate that a phenotype is due to a single gene, and chromosome aberrations that lead to characteristic alterations in the patterns of spore viability can be detected.

TABLE 1. Unified nomenclature for *C. cinereus* chromosomes

Chromosome	Size	Japanese map (J)	British map (B)	Genetic linkage	Physical linkage
I	5.1 mb	I	I	*A* factors	*A43* clone
II	3.6 mb	II?	II?	*B* factors	
III	3.6 mb	VI?	IV?		
IV	3.4 mb	III	III	*HIS4* (J)	
				TRP1 (B)	*TRP1* clone
V	3.2 mb	IV	G	*TRP1* (J)*	
				TRP2 (B)*	*TRP2* clone
VI	3.0 mb	?	J		*rDNA* clone
VII	2.7 mb	?	H		*ACU7* clone
VIII	2.6 mb	V?	V?		
IX	2.4 mb	VII?	VI?		
X	2.3 mb	?	VII?		
XI	2.2 mb	?	?		
XII	2.2 mb	?	?		
XIII	1.2 mb	?	?		RFLP

* These two genes fail to complement (unpublished observations)

Moore (1966) developed a simple method to dry gill segments, although he preferred to scatter the tetrads onto the glass before isolation. In our hands, gills are most easily prepared for drying when stripped from fruiting bodies that have just begun to shed spores (at the start of the final light period). We use forceps to split the double-sided gill into two layers, and lay the "half-gill" segment spore-side up at one end of a sterile 1.5" X 3" microscope slide. However, since it was difficult for us to apply Moore's scattering method consistently, we have returned to using a micromanipulator to remove the tetrads from the dried gill segments. The gills are sufficiently dry within an hour. If the air is quite humid, the slide can be supported above a layer of silica gel in a Petri dish to facilitate drying and storage. It is best that the slide not contact the silica particles, since these tend to cling to the glass and might damage the microscope. The slides remain usable for at least several weeks if kept dry.

We use a Jena Ergaval microscope equipped with a long working distance condenser, 16X eyepieces, and a 10X achromat objective. The gills are dried onto one end of a 1.5" X 3" microscope slide, and 10 slabs of dissection agar, each about 0.5 cm^2 are placed at the other end of the slide. The dissection agar is prepared by allowing 10 ml of supplemented YMG medium with 4% agar to solidify in a standard Petri dish, and the slabs are cut with a sterile spatula. Spore germination is enhanced by including supplements in the dissection agar. We have not noticed any consistent benefits from including furfuraldehyde (Emerson, 1954) in the plates. The slide is inverted and held 2 cm above the microscope stage using an aluminum U-shaped chamber. The open side faces a Jena micromanipulator. We form glass needles from 3mm soft glass tubing. The needle is clamped into the micromanipulator, and used to remove a tetrad of basidiospores from the dried gill and to place the tetrad in the center of the agar block. There is some loss of optical quality, since the tetrads are viewed through the slide, but with a bit of practice, the method becomes quite routine.

The micromanipulator can also be used to separate the 4 spores, but we find it is faster to do this free hand. The slide is viewed with a Wild M3 dissecting microscope using plan 15X oculars (60X magnification) and the spores spread to the edge of the field of view. The spores are allowed to germinate overnight at 37°C, and transferred to plates using a tungsten needle for subsequent genetic analysis.

We are constructing isogenic strains that differ only at the mating type loci (required for mating compatibility and fruit body formation). These strains have the 218 karyotype and either the *A3B1* mating haplotypes or the *A43B43* haplotypes. According to the calculations of Leslie (1981), our strains are 12% allogenic on all chromosomes (except for chromosomes I and II which contain the mating type loci) after five backcrosses. Initially, spore viability was quite low (25%), presumably due to multiple translocation heterozygosities in the starting strains. However, spore viability increased to 95% after only three backcrosses, and has remained high.

A particular advantage of tetrad analysis is that it allows rapid confirmation that a particular trait is caused by a single gene. For example, in the course of random spore analysis, it is common to observe that germination rates vary, and colony size can show considerable variability. Our tetrad analysis showed that in our strains, two spores always germinated and grew more slowly than the remaining two, and furthermore, this *sl*ow germination allele (*slg*$^-$) was linked to the *A3* mating type factor. We improved the 218 strain by selecting a recombinant with the normal germination rate linked to the *A3* mating type factor (*SLG*+ *A3*). Since the *A43* mating type factor was already linked to *SLG*+, we were able to eliminate this potentially deleterious gene from the isogenic strains.

Two other unfavorable variants were removed by a similar process. We noted that strain 218 harbored a mutation that we presume was spontaneous that blocks *ba*sidiospore *d*evelopment (*bad*-). Tetrad analysis confirmed that the sporeless phenotype was controlled by this single recessive gene. Furthermore, we noted recently that a second gene blocks the ability of spores to *d*arken a*n*d *d*rop (*dnd*-) during cap autolysis. Our starting strain also harbored a recessive mutation in this gene (*dnd*-). Homozygous *dnd*- strains produce brown spores that are not projected from the autolyzing fruiting body. Continued elimination of genes that confer undesirable traits when homozygous should allow continued vigor of the isogenic strains.

3.3. Physical Methods

It is time-consuming to work with multiple marked strains, but fortunately, several molecular methods can be used to facilitate mapping studies. First, it is now possible to use gel electrophoresis to resolve chromosomal DNAs of *C. cinereus*, since the length of each chromosome is distinct (Pukkila & Casselton, 1991). If DNA molecules are forced to change their orientation in response to changing electric fields, even large DNAs can be resolved. Both OFAGE (orthogonal field alternation gel electrophoresis; Carle & Olson, 1984) and CHEF (contour-clamped homogeneous electric field; Chu *et al.*, 1986) methods are suitable for *C. cinereus* chromosomes, and the required apparatus can either be assembled or purchased.

Both methods can be used to estimate the sizes of *C cinereus* chromosomes. For our standard strain 218, most of the chromosomes migrate at a position intermediate to those of *Schizosaccharomyces pombe* and *Saccharomyces cerevisiae*, allowing estimates of the largest and smallest chromosome sizes. To more accurately estimate the intermediate sizes, we took advantage of a spontaneous variant of strain 218 that exhibited an altered karyotype. Chromosome VI migrated in the position of chromosome VIII in the variant strain, designated 218-7. This change appears to have been caused by the loss of 40 tandemly repeated copies of the genes encoding the ribosomal RNAs. Changes in the rDNA copy number were measured by digesting with restriction enzymes that fail to cut within the rDNA repeat, and probing with the cloned rDNA sequences. We concluded that chromosome VI is 0.4 mb smaller in 218-7 than in 218, and thus the difference in size between chromosomes VI and VIII should also be 0.4 mb, as reflected in Table 1. These methods have also revealed surprising differences in chromosome length from strain to strain. In addition to changes in rDNA copy number, these chromosome length polymorphisms (CLPs) appear to involve translocations, and also variations in the amount of DNA that is in fact dispensable in this organism. Thus, particular chromosomes cannot be recognized in various strains simply on the basis of size. We have used cloned probes to assign mapped genes (and hence linkage groups) to chromosomes form both strain 218 (Table 1) as well as other strains (Pukkila & Casselton, 1991). We have also assigned previously unmapped genes and DNA probes to chromosomes (P.J. Pukkila and C. Skrzynia, unpublished; M. Zolan, personal communication).

Restriction fragment length polymorphism (RFLP) analysis has also been applied to facilitate gene mapping (Cassidy *et al.*, 1984; M. Zolan, personal communication). This method relies on chance substitutions that lead to the gain or loss of particular restriction enzyme recognitions sites (Botstein *et al.*, 1980). Application of the method to genetic analysis of filamentous fungi has been described in detail by Metzenberg *et al.* (1984). It is likely to prove extremely powerful for *C. cinereus*, because of the high level of natural polymorphism, and because these markers have no deleterious effects on fruiting ability.

The drawback of the RFLP method is that it requires the use of cloned probes. A variant of this method that relies on chance amplification of short segments of the genome using the polymerase chain reaction (PCR) procedure was published recently (Williams *et al.*, 1990). These randomly amplified polymorphic DNA (RAPD) markers could also be used to facilitate the mapping of new genes when use of conventionally marked strains is not desirable.

4. DNA-MEDIATED TRANSFORMATION

By far the most powerful tool of genetic analysis available for *C. cinereus* that sets it apart from other agarics is the availability of a simple transformation system (Binninger *et al.*, 1987). This system has been applied to the isolation and analysis of many genes (reviewed in Pukkila & Casselton, 1991). Using cosmid libraries, about 1 in 10^4 viable protoplasts become transformed, although some strains transform less efficiently. In our cosmid library constructed in LLC 5200, the vector (10 kb) includes the selectable *C. cinereus TRP1* gene, and the average insert of genomic DNA is 38 kb. Thus, on average, 1 transformant in 1000 should have the particular gene of interest. As long as reversion of the mutation under study is not too frequent, virtually any gene with a phenotype that is not too difficult to assay can be cloned. Typically, particular pools of the cosmid library are used for the initial transformations, so that the cosmid of interest can be identified quickly by additional transformations using subsets of the pool that was positive (a process called sib-selection).

Among the most interesting genes whose isolation and analysis depended on the existence of the transformation system is the *A* mating type factor (Mutasa *et al.*, 1990; May *et al.*, 1991). Mutasa and colleagues were able to "walk" from the nearby *PAB1* gene to clone the *A42* factor from strain Java 6. Little or no hybridization of these sequences to two other *A* haplotypes were detected. May and colleagues recovered the *A43* factor by sib-selection. They found that the factor was more complex than had been documented by previous genetic studies. The early work predicted that two genes would be present, but instead, three genes, called *A(a), A(b), and A(c)*, each capable of independent function following transformation were found, and the presence of additional genes in the region was not excluded. They also mapped a deletion found in the constitutive variant, *Amut*. Taken together, the results supported an activator model of *A* function. They observed considerable cross-hybridization with each *A43* gene to strains with altered haplotypes in the *A* region. Surprisingly, these genes appeared to be present in different combinations in different haplotypes, indicating either that recombination in nature is more frequent than has been measured in the laboratory, or that the recombinants have selective value.

The mating type work is of potential practical benefit, since strains with a variety of *A* haplotypes appear to express functions controlled by *A* following introduction of a single *A* gene by transformation. If the cloned *B* factor behaves in a similar fashion, a transforming vector could be designed that would allow most monokaryotic strains, either laboratory strains or those newly isolated from the wild, to fruit. This would be a particularly efficient method to screen for useful variants that affect fruit body development in *C. cinereus*.

It is also possible to manipulate the *C. cinereus* genome, since the incoming DNA integrates homologously in a fraction (about 5%) of the transformants. Somewhat unexpectedly, DNA form and size did not appear to effect this frequency of targeted transformation (Binninger *et al.*, 1991). It was found that single stranded DNA could be used to reduce the recovery of transformants carrying multiple tandem insertions of the transforming DNA. Targeted modifications of the *TRP1* locus have been recovered (T. Freedman, C. Skrzynia & P.J. Pukkila, unpublished observations). One potential difficulty with this type of alteration is that simple insertion of circular transforming DNA into a homologous region leads to a duplication of genomic sequences separated by the single copy vector. In some fungi, such duplications have shown to be unstable and subject to extensive methylation (Goyon & Faugeron, 1989; Faugeron *et al.*, 1990) or both methylation and point mutation (Selker *et al.*, 1987; Cambareri *et al.*, 1989). Although the nuclear genome of *C. cinereus* is methylated (Zolan & Pukkila, 1986) and *de novo* methylation of transforming DNA has been observed (T. Freedman & P.J. Pukkila, unpublished observations), duplications in *C. cinereus* appear to be stable.

ACKNOWLEDGMENTS

I thank the colleagues who provided strains, probes, and unpublished results. The work from my laboratory that is described in this chapter was supported by the ACS, NIH, and NSF.

REFERENCES

BINNINGER, D.M., SKRZYNIA, C., PUKKILA, P.J. & CASSELTON, L.A. (1987). DNA-mediated transformation of the basidiomycete *Coprinus cinereus*. *EMBO Journal* **6**, 835-840.

BINNINGER, D.M., LE CHEVANTON, L., SKRZYNIA, C., SHUBKIN, C.D. & PUKKILA, P.J. (1991). Targeted transformation in *Coprinus cinereus*. *Molecular and General Genetics* **227**, 245-251.

BOTSTEIN, D., WHITE, R.L., SKOLNICK, M. & DAVIS, R.W. (1980). Construction of a genetic linkage map using restriction fragment length polymorphisms. *American Journal of Human Genetics* **32**, 314-331.

CAMBARERI, E.B., JENSEN, B.C., SCHABTACH, E. & SELKER, E.U. (1989). Repeat-induced G-C to A-T mutations in *Neurospora. Science (Washington, D.C.)* **244**, 1571-1575.

CARLE, G.F. & OLSON, M.V. (1984). Separation of chromosomal DNA molecules from yeast by orthogonal-field-alternation gel electrophoresis. *Nucleic Acids Research* **12**, 5647-5664.

CASSELTON, L.A. & CASSELTON, P.J. (1974). Functional aspects of fluoroacetate resistance in *Coprinus* with special reference to acetyl-CoA synthetase activity. *Molecular and General Genetics* **132**, 255-264.

CASSIDY, J.R., MOORE, D., LU, B.C. & PUKKILA, P.J. (1984). Unusual organization and lack of recombination in the ribosomal RNA genes of *Coprinus cinereus. Current Genetics* **8**, 607-613.

CHU, G., VOLLRATH, D. & DAVIS, R.W. (1986). Separation of large DNA molecules by contour-clamped homogeneous electric fields. *Science (Washington, D.C.)* **234**, 1582-1585.

DUTTA, S.K. (1974). Repeated DNA sequences in fungi. *Nucleic Acids Research* **11**, 1411-1419.

EMERSON, M.R. (1954). Some physiological characteristics of ascospore activation in *Neurospora crassa. Plant Physiology* **29**, 418-428.

FAUGERON, G., RHOUNIM, L. & ROSSIGNOL, J-L. (1990). How does the cell count the number of ectopic copies of a gene in the premeiotic inactivation process acting in *Ascobolus immersus*? *Genetics* **124**, 585-591.

FRIES, N. (1947). Experiments with different methods of isolating physiological mutations of filamentous fungi. *Nature* **159**, 199.

GIBBINS, A.M.V. & LU, B.C. (1982). An ameiotic mutant of *Coprinus cinereus* halted prior to pre-meiotic S-phase. *Current Genetics* **5**, 119-126.

GOYON, C. & FAUGERON, G. (1989). Targeted transformation of *Ascobolus immersus* and de novo methylation of the resulting duplicated DNA sequences. *Molecular and Cellular Biology* **9**, 2818-2827.

HOLM, P.B., RASMUSSEN, S.W., ZICKLER, D., LU, B.C. & SAGE, J. (1981). Chromosome pairing, recombination nodules and chiasma formation in the basidiomycete *Coprinus cinereus. Carlesberg Research*

Communications **46**, 305-346.

KAMADA, T., KATSUDA, H. & TAKEMARU, T. (1984). Temperature-sensitive mutants of *Coprinus cinereus* defective in hyphal growth and stipe elongation. *Current Microbiology* **11**, 309-312.

KAMADA, T., SUMIYOSHI, T., SHINDO, Y. & TAKEMARU, T. (1989*a*). Isolation and genetic analysis of resistant mutants to the benzimidazole fungicide benomyl in *Coprinus cinereus*. *Current Microbiology* **18**, 215-218.

KAMADA, T., SUMIYOSHI, T. & TAKEMARU, T. (1989*b*). Mutations in β-tubulin block transhyphal migration of nuclei in dikaryosis in the homobasidiomycete *Coprinus cinereus*. *Plant Cell Physiology* **30**, 1073-1080.

KAMADA, T., HIRAMI, H., SUMIYOSHI, T., TANABE, S. & TAKEMARU, T. (1990). Extragenic suppressor mutations of a β-tubulin mutation in the basidiomycete *Coprinus cinereus*: isolation and genetic and biochemical analyses. *Current Microbiology* **20**, 223-228.

KANDA, T. & ISHIKAWA, T. (1986). Isolation of recessive developmental mutants in *Coprinus cinereus*. *Journal of General and Applied Microbiology* **32**, 541-543.

KANDA, T., GOTO, A., SAWA, K., ARAKAWA, H., YASUDA, Y. & TAKEMARU, T. (1989). Isolation and characterization of recessive sporeless mutants in the basidiomycete *Coprinus cinereus*. *Molecular and General Genetics* **216**, 526-529.

KANDA, T., ARAKAWA, H., YASUDA, Y. & TAKEMARU, T. (1990). Basidiospore formation in a mutant of the incompatibility factors and mutants that arrest at meta-anaphase I in *Coprinus cinereus*. *Experimental Mycology* **14**, 218-226.

LESLIE, J.F. (1981). Inbreeding for isogeneity by backcrossing to a fixed parent in haploid and diploid eukaryotes. *Genetical Research (Cambridge)* **37**, 239-252.

MAY, G., LE CHEVANTON, L. & PUKKILA, P.J. (1991). Molecular analysis of *Coprinus cinereus* mating type *A* factor demonstrates an unexpectedly complex structure. *Genetics* **128**, 529-538.

METZENBERG, R.L., STEVENS, J.N., SELKER, E.U. & MORZYCKA-WROBLEWSKA, E. (1984). A method for finding the genetic map position of cloned DNA fragments. *Neurospora Newsletter* **31**, 35-39.

MOORE, D. (1966). New method of isolating the tetrads of agarics. *Nature*

209, 1157-1158.

MOORE, D. & PUKKILA, P.J. (1985). *Coprinus cinereus*: An ideal organism for studies of genetic and developmental biology. *Journal of Biological Education* **19**, 31-40.

MUTASA, E.S., TYMON, A.M., GOTTGENS, B., MELLON, F.M., LITTLE, P.F.R. & CASSELTON, L.A. (1990). Molecular organisation of an *A* mating type factor of the basidiomycete fungus *Coprinus cinereus*. *Current Genetics* **18**, 223-229.

NORTH, J. (1982). A dominance modifier for cycloheximide resistance in *Coprinus cinereus*. *Journal of General Microbiology* **128**, 2747-2753.

NORTH, J. (1987). Linkage map of *Coprinus cinereus* (Schaeff. ex Fr.)S. F. Gray, Ink cap. In *Genetic Maps 1987 Volume 4*, pp. 340-345. Edited by S. J. O'Brian. Cold Spring Harbor: Cold Spring Harbor Laboratory Press.

PUKKILA, P.J. & CASSELTON, L.A. (1991). Molecular genetics of the agaric *Coprinus cinereus*. In *More Gene Manipulations in Fungi*, pp. 126-150. Edited by J.W. Bennett & L.L. Lasure. San Diego: Academic Press.

PUKKILA, P.J. & LU, B.C. (1985). Silver staining of meiotic chromosomes in the fungus, *Coprinus cinereus*. *Chromosoma (Berlin)* **91**, 108-112.

RAO, P.S. & NIEDERPRUEM, D.J. (1969). Carbohydrate metabolism during morphogenesis of *Coprinus lagopus* (*sensu* Buller). *Journal of Bacteriology* **100**, 1222-1228.

SELKER, E.U., CAMBARERI, E.B., JENSEN, B.C. & HAACK, K.R. (1987). Rearrangement of duplicated DNA in specialized cells of *Neurospora*. *Cell* **51**, 741-752.

SWAMY, S., UNO, I. & ISHIKAWA, T. (1984). Morphogenetic effects of mutations at the *A* and *B* incompatibility factors in *Coprinus cinereus*. *Journal of General Microbiology* **130**, 3219-3224.

TAKEMARU, T. (1982). Genetic map of *Coprinus macrorhizus*. In *Experimental Methods in Microbial Genetics*, pp. 355-356. Edited by T. Ishikawa. Tokyo: Kyoritsu Publishing Company, Ltd.

TAKEMARU, T. & KAMADA, T. (1972). Basidiocarp development in *Coprinus macrorhizus*. I. Induction of developmental variations. *Botanical Magazine of Tokyo* **85**, 51-57.

VEAL, D. & CASSELTON, L.A. (1985). Regulation of tryptophan metabolism in *Coprinus cinereus*: isolation and characterisation of mutants resistant to 5-fluoroindole. *Archives of Microbiology* **142**, 157-163.

WILLIAMS, J.G.K., KUBELIK, A.R., LIVAK, K.J., RAFALSKI, J.A. & TINGEY, S.V. (1990). DNA polymorphisms amplified by arbitrary primers are useful as genetic markers. *Nucleic Acids Research* **18**, 6531-6535.

ZOLAN, M.E. & PUKKILA, P.J. (1986). Inheritance of DNA methylation in *Coprinus cinereus*. *Molecular and Cellular Biology* **6**, 195-200.

ZOLAN, M.E., TREMEL, C.J. & PUKKILA, P.J. (1988). Production and characterization of radiation-sensitive meiotic mutants of *Coprinus cinereus*. *Genetics* **120**, 379-387.

CHAPTER 13

APPLICATION OF ARBITRARILY-PRIMED POLYMERASE CHAIN REACTION IN MOLECULAR STUDIES OF MUSHROOM SPECIES WITH EMPHASIS ON *LENTINULA EDODES*

Siu-wai Chiu[1], Hoi-shan Kwan[1], and Suk-chun Cheng[2]

[1]Department of Biology, The Chinese University of Hong Kong, Shatin, New Territories, Hong Kong.

[2]Department of Applied Biology and Chemical Technology, Hong Kong Polytechnic, Kowloon, Hong Kong.

1. INTRODUCTION

Most studies on mushrooms are about cultivation methods. In contrast, genetic information on most mushroom species is scanty, except in the case of *Agaricus bisporus* (syn. *A. brunnescens*) (see Anderson, Chapter 10). Some morphological mutants of *Lentinula edodes* were isolated and characterized (Hasebe *et al.*, 1982, 1987, 199; Murakami *et al.*, 1987; Itavaara, 1990). However, these mutants usually do not have any commercial value. For wild and cultivated strains of mushroom species, strain characterization/typing has traditionally been based on morphological characters and physiological properties such as temperature optima. Recently, the introduction of biochemical markers, such as soluble protein profiles, various enzyme activities, serological properties (Kawamura & Goto, 1980; Ohmasa & Furukawa, 1986; Itavaara, 1988; Burdsall *et al.*, 1990; Lin & Hsieh, 1991) and especially, isozyme profiles (see Royse & May, Chapter 11), represent a significant improvement in typing methodology since these techniques offer greater diversity. However, expression of these biochemical markers is still under physiological and genetic controls. Markers based upon DNA probes, on the other hand, are direct assessments of the constant genetic diversity of different strains and are dominant characters. DNA fingerprinting techniques have the potential to reveal an almost unlimited number of polymorphisms. One of the earlier methods is the widely used restriction fragment length polymorphisms (RFLPs) technique which uses specific DNA probes to identify single or low copy sequences in genomic DNA and shows allelic relationship among individuals by the presence or absence of restriction endonuclease sites in the examined sequence. Instead of using RFLPs, we initiated a molecular project on *Lentinula edodes* utilizing the recently established molecular technique, polymerase chain reaction (PCR), which was awarded ''the molecule of the year'' in 1989 (Guyer & Koshland, 1989).

Polymerase chain reaction is an *in vitro* enzymatic amplification of a specific DNA sequence, originally present in minute quantity, at an exponential rate (Arnheim *et al.*, 1990; Innis *et al.*, 1990; Mullis, 1990). Polymorphisms are sought in the distance between two target sequences tagged by two short complementary oligodeoxynucleotides called primers. A cycle of PCR consists of 3 steps:

1. Denaturation of a double-stranded DNA into two single-stranded DNA templates by a high temperature, usually 95°C,
2. Annealing of the single-stranded DNA template with specific primers at a lower temperature, and
3. Addition of DNA polymerase and amplification of the DNA template marked by the primers in the process called primer extension. This final step is dependent upon the right orientation of the primers.

The amplified DNA fragments can be revealed by electrophoresis on an agarose or a polyacrylamide gel, or characterized by direct DNA sequencing.

With the recent advances in methodology and instrumentation, PCR has dramatically impacted molecular studies in many fields of biological science (Erlich *et al.*, 1991). In fungi, PCR has been applied to the examination of ribosomal DNA sequences for molecular systematics and population studies (Bruns *et al.*, 1990; Forster *et al.*, 1990; Illingworth *et al.*, 1991). This conventional PCR requires prior information on target gene sequences and thus imposes considerable restriction on usage with mushroom species about which genetic information in general is scanty. We have used the conserved internally transcribed spacer sequence of the ribosomal DNA gene repeat as the specific primer and examined various *Lentinula edodes* and *Coprinus cinereus* strains by PCR. We found that polymorphisms among different strains of the same species could not be revealed by mobility in gel electrophoresis but were demonstrable by DNA sequencing or restriction digestion (H.S. Kwan, S.W. Chiu, K.M. Pang, S.C. Cheng, unpublished results). Therefore, conventional PCR using specific primer is not a straightforward method for strain characterization.

On the other hand, PCR using a non-specific primer has recently been developed to reveal polymorphisms in the DNA profiles of bacteria, plant species, *Neurospora crassa* and *Drosophila* (Welsh & McClelland, 1990; Wesley *et al.*, 1990; Williams *et al.*, 1990; Weining & Langridge, 1991). These PCR-based DNA fingerprinting techniques relieve the requirement of any prior sequence information and thus, reduce the cost and time taken in molecular studies. These methods also apply low stringency conditions for annealing the primer with the DNA template. This can be achieved by annealing an arbitrary primer with the template at low temperatures using a primer with degenerate sequence, a short primer or a specific primer sequence which can hybridize to various locations on the genomic DNA. In the method called random amplified polymorphic DNA (RAPD), an oligodeoxynucleotide primer which contains no palindromic sequence and possesses 50-80% G+C composition is used (Williams *et al.*, 1990). The technique has recently been applied in studies with *Agaricus bisporus* (see Anderson, Chapter 10) and *Armillaria* (Smith & Anderson, 1991). In the method called arbitrarily-primed polymerase chain reaction (AP-PCR), a primer with an arbitrary sequence is randomly chosen. In this method, by adopting a lower temperature to anneal the arbitrary primer with the template DNA, various regions of the genomic DNA are tagged with the arbitrary primer and so amplified by added DNA polymerase (Welsh & McClelland, 1990). We describe here the first study utilizing AP-PCR for typing mushroom strains and explore its possible application to mushroom science.

2. EXPERIMENTAL PROCEDURES

2.1. Mushroom Strains

Both cultivated and wild strains of *Lentinula edodes* and their single spore isolates were kindly provided by Prof. S. T. Chang and maintained on potato dextrose agar/broth at 25°C in darkness. *Coprinus cinereus*, *Volvariella bombycina* and *V. volvacea* strains were maintained on potato dextrose agar/broth supplemented with 0.2% yeast extract and incubated at 30°C in darkness.

2.2 Dedikaryotization

The method developed by Leal-Lara & Eger-Hummel (1982) was adopted. A 7-day old dikaryotic plate culture was homogenized using a Waring blender. The homogenate was pipetted into a 50ml conical flask containing 10ml of the dedikaryotization solution [glycine (Merck, Serva), 5 g/l; glucose monohydrate (Merck 8324), 20 g/l] at a final concentration of 20 hyphal fragments per ml. It took 4 weeks for the hyphal fragments to grow into visible balls of about 3 mm in diameter in stationary cultures incubated at 25°C. This culture was homogenized once more with 10 ml of water. The resultant hyphal fragments were serially diluted and plated onto potato dextrose agar. After incubation for 1 week at 25°C, visible colonies were picked and transferred to a fresh agar plate. Dedikaryotized isolates were selected by the absence of clamp connections, and their mating types were determined.

2.3. DNA Preparation

A mini-preparation of total DNAs (Lee *et al.*, 1988) was performed as follows: 0.2 to 1 g mycelium was deep-frozen by liquid nitrogen, ground to a fine powder using a mortar and pestle and incubated in 0.4 ml lysis buffer (1% 2-mercaptoethanol, 3% SDS and 50 mM EDTA in 50 mM Tris-HCl buffer, pH 7.2) at 65°C for 1 hr. Cell debris was then removed by centrifugation and RNase added to the supernatant to a concentration of 50 μg/ml. After incubation at 37°C for 1 hr, the reaction mixture was extracted with phenol and chloroform and the DNA precipitated with ethanol. The concentration and purity of a DNA sample were determined by agarose gel electrophoresis and ethidium bromide staining.

2.4. Arbitrarily-primed Polymerase Chain Reaction (AP-PCR)

Polymerase chain reaction is especially sensitive to contamination. Therefore, great care and cleanliness are essential. Arbitrarily primed-PCR was described by Welsh and McClelland (1990). A 10 µl reaction mixture contained 0.025 units of AmpliTaq polymerase, 1X buffer and 0.2 mM each dNTP (Perkin Elmer Cetus), 4 mM $MgCl_2$, 10 µM primer (Operon Technologies, Alameda) and 5 ng template DNA. The four arbitrary primers used in this study were:

1. M13 forward sequencing primer (-47),
   5'd-CGCCAGGGTTTTCCCAGTCACGAC-3';
2. M13 reverse sequencing primer (-48),
   5'd-AGCGGATAACAATTTCACACAGGA-3';
3. *Eco*RI-Ext primer,
   5'd-TAGGCGTATCACGAGGCCCT-3'; and
4. Gal K primer,
   5'd-TACGGTGGCGGAGCGCAGCA-3'.

The mixture was overlaid with oil and subjected to two cycles of low stringency amplification: 94°C, 3 min for denaturing template DNA; 35°C, 5 min for low stringency annealing of primer and 72°C, 5 min for primer extension. Ten high stringency cycles then followed: 94°C, 1 min; 50°C, 1 min and 72°C, 2 min. At the end of these cycles, 90 µl of solution containing 2.25 units of AmpliTaq polymerase, 1X buffer and 0.2 mM each dNTP was added to the sample. The reaction mixture was subjected to another 40 cycles of high stringency amplification: 94°C, 0.5 min (ramp time = 1 min); 50°C, 1 min; 72°C, 1.5 min. The products of the reaction were revealed by agarose gel electrophoresis and ethidium bromide staining. Best resolution of the amplified DNA bands was obtained when the products were run on a 3% agarose gel [2% NuSieve agarose and 1% SeaKem agarose (FMC)]. Also, to confirm that the observed bands were amplified genomic DNA and not primer artifacts, genomic DNA was omitted from control reactions for each primer.

3. PARAMETERS AFFECTING AP-PCR PATTERNS

Several parameters were tested to determine their effects on the amplified

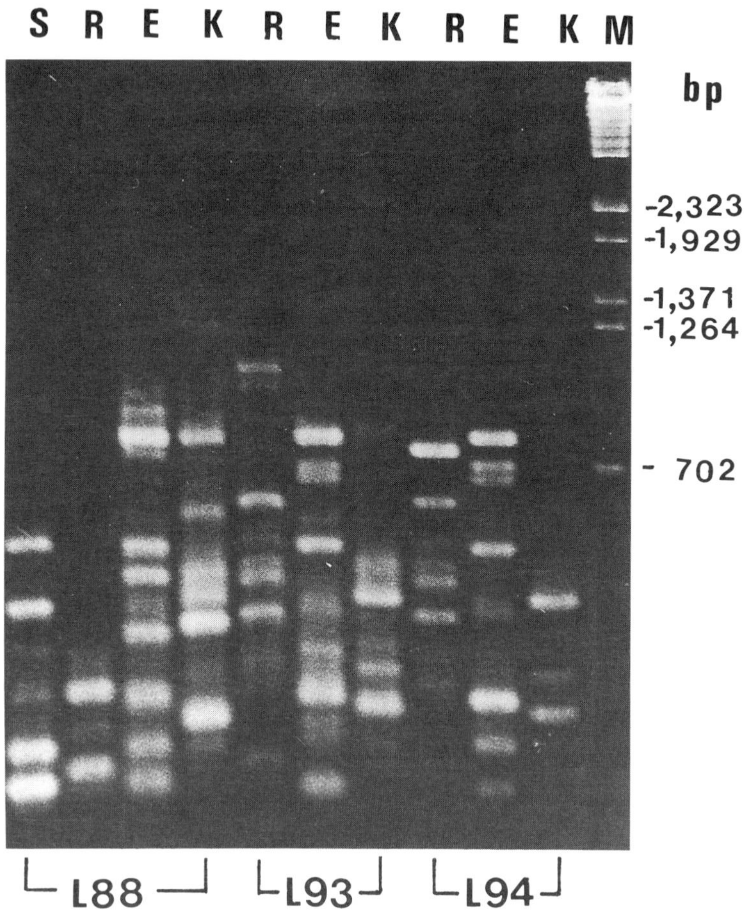

FIGURE 1. Effect of different arbitrary primers on the PCR-amplified DNA profiles for three *Lentinula edodes* strains. Arbitrary primers: S, M13 forward sequencing primer; R, M13 reverse sequencing primer; E, *Eco*RI-Ext primer; K, Gal K primer. L, *L. edodes* strains. M, molecular weight standard (*Bst* EII-digested phage lambda DNA).

DNA profiles.

3.1. Choice of Arbitrary Primer

Two universal primers which are used for DNA sequencing and two primers synthesized for a bacterial genetics project were randomly chosen and used as the arbitrary primers. Three different monokaryotic *Lentinula edodes* strains were examined with these arbitrary primers (Figure 1). For every arbitrary primer, the three strains showed strain-specific DNA fingerprints. For every strain, different primers amplified different genomic regions, and thus showed distinct DNA profiles.

3.2. Type of DNA Polymerase and Its Concentration

Originally, PCR utilized the Klenow fragment of *Escherichia coli* DNA polymerase I for primer extension. However, this enzyme is inactivated at the high temperature used for denaturing the double-stranded DNA. As a result, it was necessary to add DNA polymerase after every amplification cycle. The introduction of thermostable DNA polymerases, such as Taq DNA polymerase produced by the thermophilic bacterium *Thermus aquaticus,* or its recombinant form AmpliTaq DNA polymerase which was cloned in *Escherichia coli,* not only simplifies the procedure for PCR but also improves the specificity and yield of the amplification cycle (Erlich *et al.*, 1991). The commonly used Taq and AmpliTaq DNA polymerases (Perkin Elmer Cetus) do not possess 3' —> 5' exonuclease activity. On the other hand, Vent DNA polymerase (New England Biolabs), which is synthesized by the extreme thermophile *Thermococcus litoralis,* possesses 3' —> 5' exonuclease activity. As a result, the latter enzyme has a lower misincorporation rate. When Vent DNA polymerase was used in AP-PCR studies, it was as equally effective as AmpliTaq DNA polymerase (Figure 2). However, the amplified DNA profiles were different when the different DNA polymerases were used; fewer amplified DNA bands were revealed with Vent DNA polymerase. In addition, if too much Vent was used, a smear rather than discrete DNA bands resulted on subsequent gel electrophoresis.

3.3. Reproducibility of AP-PCR Patterns of Different DNA Preparations of the Same Strain

It is our experience that when the DNA preparation was brown in color, no amplification occurred. In addition, when the concentration of template DNA was low, very few bands were amplified. Therefore, a preliminary running

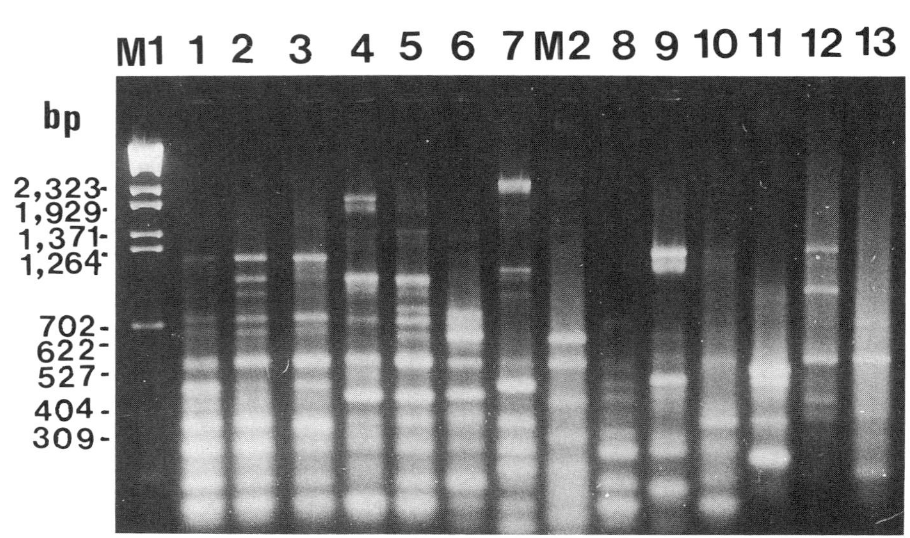

FIGURE 2. Effect of type of DNA polymerase and its concentration on the PCR-amplified DNA profiles of fungal strains using M13 forward sequencing primer as the arbitrary primer. M, molecular weight standards: M1, *Bst* EII-digested phage lambda DNA; M2, *Msp* I-digested plasmid pBR322 DNA. Lanes 1 to 4, *Lentinula edodes* strains L13, L15x21, L54 and L76. Lanes 5, 8, 10, 12 & 13, *L. edodes* strain L88. Lanes 9 & 11, *Schizosaccharomyces pombe* strain ATCC24843. Lane 6, *Coprinus cinereus* strain M6. Lane 7, *Saccharomyces cerevisiae* strain Y77. All samples were amplified using the protocol as described in the Section 2.4. except lanes 12 and 13 for which the annealing temperature for the relax cycle was 40°C, and for the stringent PCR cycles, the temperature was 50°C. Also, 0.025 units of Vent was used for lane 12 and 2.5 units of Vent was used for lane 13. Amplitaq DNA polymerase was used for Lanes 1 to 9 and Vent DNA polymerase for Lanes 10-13.

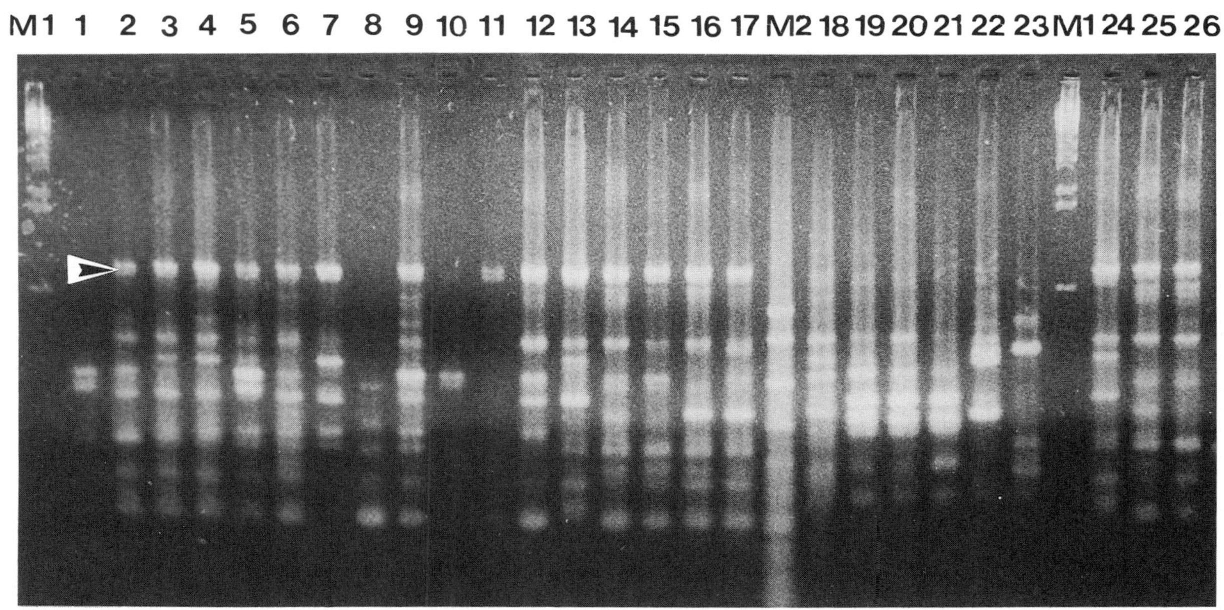

FIGURE 3. Agarose gel electrophoresis of PCR-amplified products of fungal strains using *Eco*RI-Ext primer. M, molecular weight standards: M1, *Bst* EII-digested phage lambda DNA; M2, *Msp* I-digested plasmid pBR322 DNA. All lanes are *Lentinula edodes* strains except for the following lanes: lanes 18-21, *Coprinus cinereus* strains; lane 22, *Saccharomyces cerevisiae* strain Y77; lane 23, *Schizosaccharomyces pombe* strain ATCC24843. Lanes 13-15 and lanes 24-26 are AP-PCR profiles of *L. edodes* strains L88, L93 and L94 DNAs prepared and amplified at different dates. >: common DNA band specific to *L. edodes* strains.

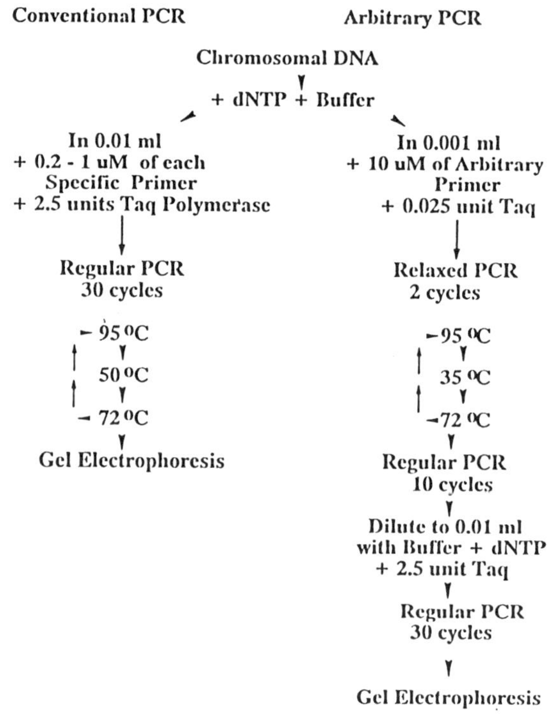

FIGURE 4. A flowchart comparing the protocols of conventional PCR using specific primer and arbitrarily-primed PCR.

of the samples on a 1.5% gel serves as a useful check on the quality of the amplified products. Figures 3 shows that many samples can be compared on the same gel. Also, DNA samples from three *Lentinula edodes* strains prepared on different dates yielded almost identical DNA profiles indicating high reproducibility of the data obtained using this technique once the PCR conditions were optimized. In addition, at least one DNA band was found to be common among all the *L. edodes* strains tested but absent in *Coprinus cinereus* strains.

Figure 4 summarizes the protocols of the conventional polymerase chain reaction using specific primer and arbitrarily-primed polymerase chain reaction. The AP-PCR protocol was applied in molecular studies of mushroom species as discussed below.

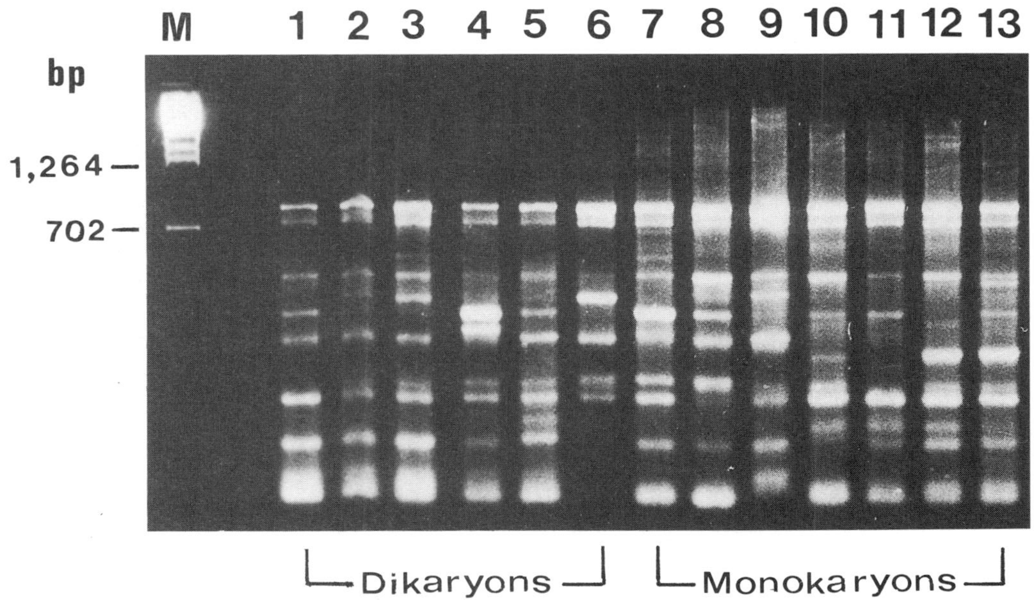

FIGURE 5. Genomic fingerprints of monokaryotic and dikaryotic *Lentinula edodes* strains by AP-PCR using *Eco*RI-Ext primer. M, molecular weight standard (*Bst*EII-digested phage lambda DNA). Lanes 1-6, dikaryotic strains L15x21, L34, L40, L45, L54 and L70. Lanes 7-13, monokaryotic strains L74, Ld5, L88, L93, L94, L95 and L96. Strains L93-96 are siblings.

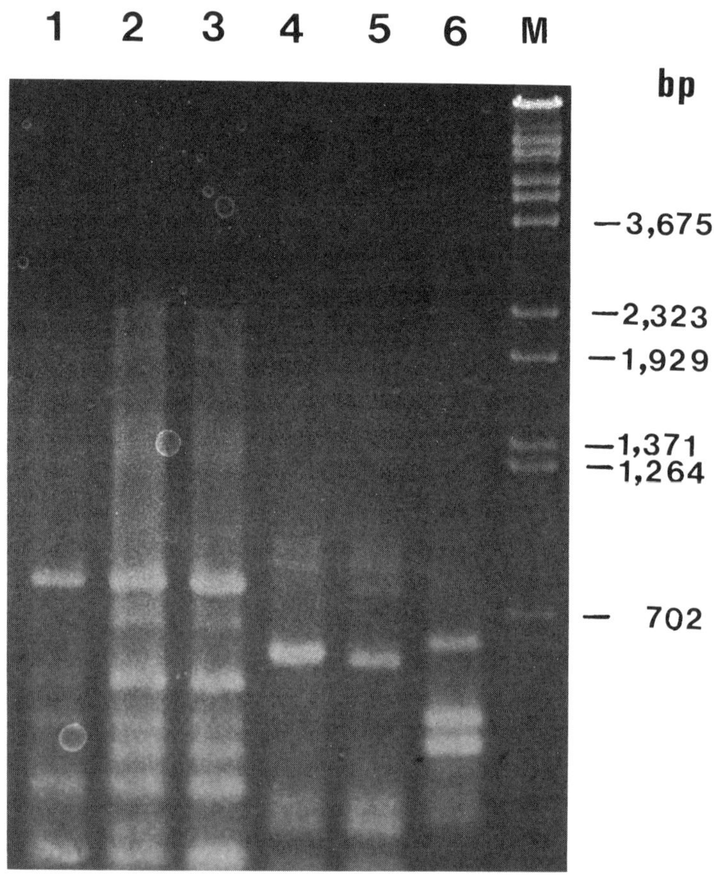

FIGURE 6. Genomic fingerprints of a putative protoplast fusion product formed between two *Volvariella* species and two dedikaryotized isolates of a *Lentinula edodes* strain by AP-PCR using *Eco*RI-Ext primer. M, molecular weight standard (*Bst* EII-digested phage lambda DNA). Lane 1, dikaryotic *L. edodes* strain L84. Lanes 2 and 3, dedikaryotized isolates Ld2 and Ld5, respectively. Lane 4, *V. bombycina* strain Vo-1. Lane 6, *V. volvacea* strain Vv34. Lane 5, putative protoplast fusion product.

4. APPLICATION OF AP-PCR

4.1. Strain Typing

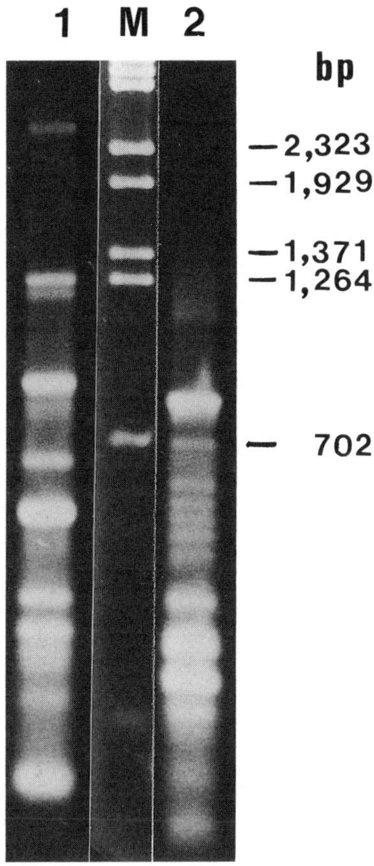

FIGURE 7. Genomic fingerprints of a putative protoplast fusion product and its parent, *Volvariella volvacea* by AP-PCR using M13 forward sequencing primer. M, molecular weight standard (*Bst* EII-digested phage lambda DNA). Lane 1, putative protoplast fusion product. Lane 2, *V. volvacea* strain Vv34.

The most important application of AP-PCR is as a means of cell line authentication.

4.1.1. Monokaryotic strains. Figure 5 shows the AP-PCR DNA profiles of 7 monokaryotic *Lentinula edodes* strains amplified with AmpliTaq using *Eco*RI-Ext primer. Each strain has a distinct DNA fingerprint. However, the

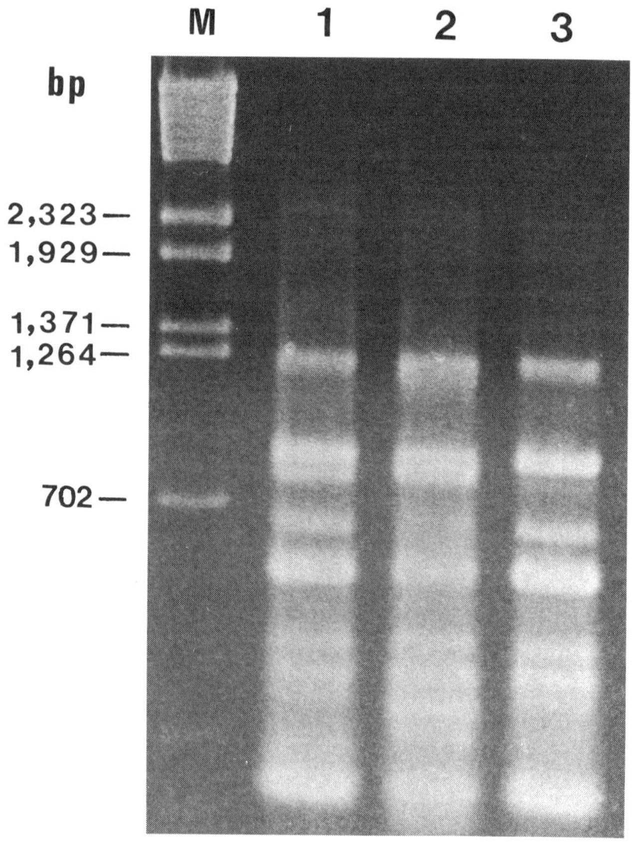

FIGURE 8. Agarose gel electrophoresis of AP-PCR amplified DNA profiles of a putative protoplast fusion product, its parent and a mixed parental DNA sample using M13 forward sequencing primer. M, molecular weight standard (*Bst*EII-digested phage lambda DNA). Lane 1 , *Volvariella bombycina* strain Vo-1. Lane 2, putative protoplast fusion product. Lane 3, mixed DNAs of *V. bombycina* strain Vo-1 and *V. volvacea* strain Vv34.

four siblings showed more common DNA bands than non-sibling strains. Polymorphic DNA bands which were present only in some strains could be identified.

4.1.2. Dikaryotic strains. Figure 5 also shows the DNA profiles of 6

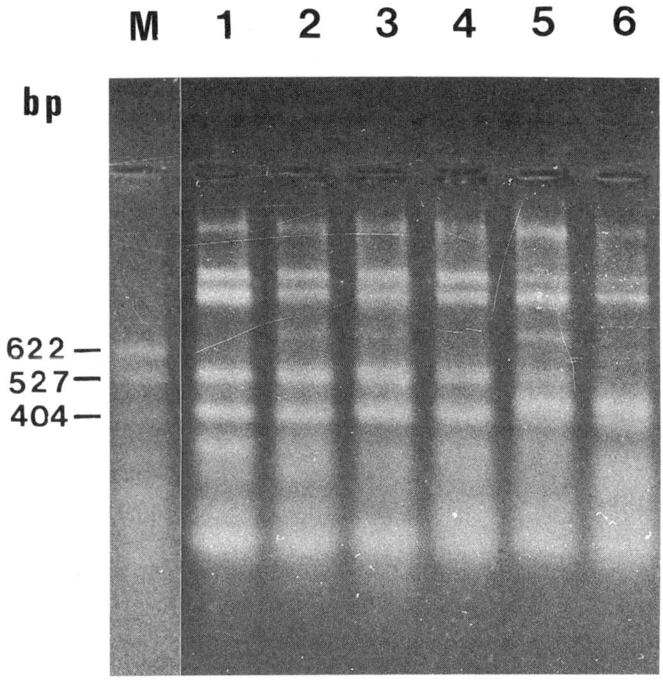

FIGURE 9. Genomic fingerprints of two synthesized *Lentinula edodes* dikaryons by AP-PCR using M13 forward sequencing primer. M, molecular weight standard (*Msp* I-digested plasmid pBR322 DNA). Lanes 1 and 3, sibling monokaryons strain L73 and L76. Lane 2, the resultant dikaryon strain L73x76. Lanes 4 and 6, non-sibling monokaryons strain L74 and L80. Lane 5, the resultant dikaryon strain L74x80.

dikaryotic *Lentinula edodes* strains. Again, each strain has a distinct DNA fingerprint. In addition, the complexity of a DNA profile has no relationship with the monokaryotic or dikaryotic nature of the strain being tested. Nevertheless, all the strains could be characterized at the DNA level by AP-PCR using just one arbitrary primer, in this case the *Eco*RI-Ext primer.

4.1.3. Characterization of a putative protoplast fusion product. Confirmation of a protoplast fusion product has been based on morphological markers, isozyme markers (see Peberdy & Fox, Chapter 7) and more rarely on molecular markers such as restriction fragment length polymorphisms (Castle *et al.*, 1988). Here, we tested the feasibility of using AP-PCR to generate molecular markers for the confirmation of a protoplast fusion product. Figure 6 shows the use of the *Eco*RI-Ext primer as the arbitrary primer for such confirmation. It is apparent that the putative fusion product has a profile identical with *Volvariella bombycina* but quite different from that of *V. volvacea*. The dissimilarity between the putative fusion product and *V. volvacea* could be confirmed when M13 forward sequencing primer was used as the arbitrary primer for AP-PCR (Figure 7). The putative fusion product and the parental *V. bombycina* strain differed only in the relative intensity of certain DNA bands while the amplified DNA profile of the putative fusion product could not be reproduced when equal amounts of genomic DNAs from *V. bombycina* and *V. volvacea* were mixed and used as the DNA template for amplification (Figure 8). Therefore, AP-PCR has potential in the characterization of protoplast fusion products.

4.2. Relationship Between Strains

4.2.1. Monokaryon-dikaryon relationship. Figure 9 shows the AP-PCR DNA profiles of two dikaryons, one prepared from sibling monokaryons and the other from non-sibling monokaryons. The sibling monokaryons had highly similar DNA profiles with one extra band in one monokaryon and an extra smear of smaller DNA fragments in the other. The synthesized dikaryon not only possessed the weak DNA band of one monokaryon but also showed the pattern of smear DNA bands characteristic of the other sibling monokaryon. In the case of the dikaryon synthesized from non-sibling monokaryons, which differed from each other by the presence or absence of one discrete DNA band, a new DNA band was evident in addition to the DNA bands common to both monokaryons. The relative band intensities in the dikaryon is not simply an additive effect of the DNA bands of the two component monokaryons. Irrespective of whether dikaryons are derived from sibling or non-sibling monokaryons, each monokaryon contributes in part to the profile of the resultant dikaryon and this profile can be used to reveal monokaryon-dikaryon relationships.

4.2.2. Dedikaryotization demonstration. Figure 6 shows the DNA profiles of a *Lentinula edodes* dikaryon and its compatible dedikaryotized isolates.

The dikaryon possessed the common DNA bands of the two dedikaryotized isolates. The two dedikaryotized isolates differed by the absence or presence of one small DNA band and another weaker DNA band in one isolate. Therefore, in addition to the absence of clamp connections and compatibility in the mating reaction, AP-PCR can be applied to the characterization of two dedikaryotized isolates.

4.3. Progeny Analysis And Identification Of Polymorphic Markers

As shown in Figure 5, some DNA bands were segregated in some sibling strains but not the others. Therefore, polymorphic bands can be generated with AP-PCR and if coupled with chi-square tests on a larger sample size of progeny, these polymorphic bands can be employed as genetic markers. It should be noted that it has not been established whether a DNA band which is common among strains shares an identical DNA sequence. Also, these polymorphic DNA bands do not imply allelic forms as with RFLPs.

5. CONCLUDING REMARKS

In summary, AP-PCR requires only mini-preparation of chromosomal DNAs such that many strains can be typed within a short period of time. Furthermore, the use of any arbitrary primer chosen at random combined with the ease and convenience of the PCR procedure, accords the AP-PCR technique great potential in molecular studies of mushroom species. We have demonstrated the value of AP-PCR in the following areas:

1. Typing, i.e., strain identification,
2. Confirmation of protoplast fusion products/hybrid,
3. Identification of species-specific DNA markers,
4. Establishing monokaryon-dikaryon relationships,
5. Demonstration of dedikaryotization,
6. Generation of polymorphic molecular markers which may be used as genetic markers in progeny analysis.

As AP-PCR assesses the whole genome, it reveals greater polymorphisms among strains than conventional PCR based on specific primers. The calculated dissimilarity index for *L. edodes* strains analyzed by 18 restriction fragment length polymorphisms (RFLPs) (Kulkarni, 1991) ranged from 0.43

to 0.90 while the dissimilarity index analyzed by isozyme studies based on 9 enzymes coded by 11 genetic loci (Royse & May, 1987) ranged from 0.63 to 1. We have obtained data with only one arbitrary primer suggesting similar dissimilarity index values for the *L. edodes* strains tested here (H.S. Kwan, S.W. Chiu, K.M. Pang, S.C. Cheng, unpublished results). Therefore, the potential of AP-PCR for differentiating strains is similar to that of RFLPs and isozymes. In addition, unlike RFLPs, it does not require hybridization to reveal polymorphisms, and unlike isozymes, it is not affected by physiological conditions.

6. ACKNOWLEDGEMENTS

This study is supported by the Croucher Foundation, Hong Kong. The authors thank Mr. G. L. Yang and Prof. S. T. Chang for the provision of the putative protoplast fusion product of the two *Volvariella* species and Prof. S. T. Chang for the provision of *Lentinula edodes* strains.

REFERENCES

ARNHEIM, N., WHITE, T. & RAINEY, W.E. (1990). Application of PCR: organismal and population biology. *BioScience* **40**, 174-182.

BRUNS, T.D., FOGEL, R. & TAYLOR, J.W. (1990). Amplification and sequencing of DNA from fungal herbarium specimens. *Mycologia* **82**, 175-184.

BURDSALL, H.H., BANIK, M. & COOKE, M.E. (1990). Serological differentiation of three species of *Armillaria* and *Lentinula edodes* by enzyme-linked immunosorbent assay using immunized chickens as a source of antibodies. *Mycologia* **82**, 415-423.

CASTLE, AJ., HORGEN, P.A. & ANDERSON, J.B. (1988). Crosses among homokaryons from commercial and wild-collected strains of the mushroom *Agaricus brunnescens* (=*A. bisporus*). *Applied and Environmental Microbiology* **54**, 1643-1648.

ERLICH, H.A., GELFAND, D. & SNINSKY, J.J. (1991). Recent advances in the polymerase chain reaction. *Science* **252**, 1643-1656.

FORSTER, H., COFFEY, M.D., ELWOOD, H. & SOGIN, M.L. (1990).

Sequence analysis of the small subunit ribosomal RNAs of three zoosporic fungi and implications for fungal evolution. *Mycologia* **82**, 306-312.

GUYER, R.L. & KOSHLAND, D.E., JR. (1989). The molecule of the year. *Science* **246**, 1543-1546.

HASEBE, K., MURAKAMI, S. & TSUNEDA, A. (1991). Cytology and genetics of a sporeless mutant of *Lentinus edodes*. *Mycologia* **83**, 354-359.

HASEBE, K., TOKIMOTO, K. & KOMATSU, M. (1982). "Dwarf" mutant of *Lentinus edodes* (Berk.) Sing. *Reports of the Tottori Mycological Institute* **20**, 113-116.

HASEBE, K., MURAKAMI, S. & KOMATSU, M. (1987). Genetic analysis of some morphological mutations in homokaryotic mycelia of *Lentinus edodes*. *Reports of the Tottori Mycological Institute* **25**, 56-61.

ILLINGWORTH, C.A., ANDREWS, J.H., BIBEAU, C. & SOGIN, M.L. (1991). Phylogenetic placement of *Athelia bombacina*, *Aureobasidium pullulans* and *Colletotrichum gloeosporioides* inferred from sequence comparisons of small-subunit RNAs. *Experimental Mycology* **15**, 65-75.

INNIS, M.A., GELFAND, D.H., SNINKY, J.J. & WHITE, T.J. (1990). *PCR Protocols: A Guide to Methods and Applications*. San Diego: Academic Press.

ITAVAARA, M. (1988). Identification of shiitake strains and some other basidiomycetes: protein profile, esterase and acid phosphatase zymograms as an aid in taxonomy. *Transactions of the British Mycological Society* **91**, 295-304.

ITAVAARA, M. (1990). Characterization of a shiitake mutant strain producing ballshaped fruiting bodies. *Physiologia Plantarum* **80**, 36-42.

KAWAMURA, N. & GOTO, M. (1980). Biochemical characteristics of the isolates of shiitake mushroom (*Lentinus edodes*). *Reports of the Tottori Mycological Institute* **18**, 217-224.

KULKARNI, R.K. (1991). DNA polymorphisms in *Lentinula edodes*, the shiitake mushroom. *Applied and Environmental Microbiology* **57**, 1735-1739.

LEAL-LARA, H. & EGER-HUMMEL, G. (1982). A monokaryotization method and its use for genetic studies in wood-rotting basidiomycetes. *Theoretical and Applied Genetics* **61**, 65-68.

LEE, S.B., MILGROOM, M.G. & TAYLOR, J.W. (1988). A rapid, high yield mini-prep method for isolation of total genomic DNA from fungi.

Fungal Genetics Newsletter **35**, 23-24.

LIN, S-J. & HSIEH, T-C. (1991). Application of API-ZYM enzyme testing system to rapid identification of strains in *Lentinus edodes* (Berk.) Sing. *Bulletin of the Taiwan Forest Research Institute, New Series* **6**, 21-26.

MULLIS, K.B. (1990). The unusual origin of the polymerase chain reaction. *Scientific American* **262**, 36-43.

MURAKAMI, S., HASEBE, K. & TSUNEDA, A. (1987). A mutant of *Lentinus edodes* forming aberrant clamp connections. *Reports of the Tottori Mycological Institute* **25**, 49-55.

OHMASA, M. & FURUKAWA, H. (1986). Analysis of esterase and malate dehydrogenase isozymes of *Lentinus edodes* by isoelectric focusing for the identification and discrimination of stocks. *Transactions of the Mycological Society (Japan)* **27**, 79-80.

ROYSE, D.J. & MAY, B. (1987). Identification of shiitake genotypes by multilocus enzyme electrophoresis: catalog of lines. *Biochemical Genetics* **25**, 705-716.

SMITH, M. L. & ANDERSON, J. B. (1991). Molecular genetic analysis of a local population of *Armillaria*. *Mycological Society of America Newsletter* **42**, 33.

WEINING, S. & LANGRIDGE, P. (1991). Identification and mapping of polymorphisms in cereals based on the polymerase chain reaction. *Theoretical and Applied Genetics* **82**, 209-216.

WELSH, J. & McCLELLAND, M. (1990). Fingerprinting genomes using PCR with arbitrary primers. *Nucleic Acids Research* **18**, 7213-7218.

WESLEY, C. S., BEN, M., KREITMAN, M., HAGAG, N. & EANES, W. F. (1990). Cloning regions of the *Drosophila* genome by microdissection of polytene chromosome DNA and PCR with nonspecific primer. *Nucleic Acids Research* **18**, 599-603.

WILLIAMS, J.G.K., KUBELIK, A.R. , LIVAK, K.J., RAFALSKI, J.A. & TINGEY, S.V. (1990). DNA polymorphisms amplified by arbitrary primers are useful as genetic markers. *Nucleic Acids Research* **18**, 6531-6535.

CHAPTER 14

A STRATEGY FOR ISOLATING MUSHROOM-INDUCING GENES IN EDIBLE BASIDIOMYCETES

Carlene A. Raper and J. Stephen Horton

Department of Microbiology & Molecular Genetics,
The University of Vermont, Burlington, Vermont 05405-0068, U.S.A.

1. INTRODUCTION

A gene that induces mushroom development has been isolated from the wood-rotting Basidiomycete *Schizophyllum commune*. We call this gene *FRT1* for the key role it plays in the development of fruiting bodies, commonly called mushrooms. *FRT1* has been characterized as a member of a family of similar genes in *S. commune*. We have found evidence suggesting that its activity is regulated by activity of the mating-type genes. It also appears to be involved in the regulation of other genes in the developmental pathway of fruiting. This group of genes may be conserved in related species of edible Basidiomycetes. An understanding of the structure of these genes and how they work in *S. commune* could guide us towards the isolation and characterization of similar genes in commercially valuable species.

2. MUSHROOM DEVELOPMENT IN *SCHIZOPHYLLUM*

S. commune has long served as a model system for the study of genes regulating mating and fruiting in fungi. While edible (and quite delicious), its mushrooms are relatively puny and chewy and therefore of no commercial value for gastronomic purposes. Nevertheless, this species is ideal for genetic studies, not only because it is haploid throughout most of its life cycle, but because it can be made to complete this cycle on chemically defined media within a relatively short period of time. (See the following for reviews of studies on mating and fruiting in *S. commune*: Raper, 1965; Wessels, 1969, 1987; Niederpruem & Wessels, 1969; Volz & Niederpruem, 1969; Schwalb, 1978; Leslie & Leonard, 1979; Raudaskoski & Vauras, 1982; Raper, 1983, 1988; Stankis *et al.*, 1990; Novotny & Stankis, 1991.)

S. commune is a heterothallic species in which mushrooms normally develop only after activation of the mating-type genes by the coupling of two homokaryons (single spore isolates) of different mating types to form a dikaryon. Our cloned *FRT1* gene overrides the normal requirement of a compatible mating interaction for fruiting in this fungus. It is capable of inducing mushroom development in certain unmated homokaryons when integrated into the genome by DNA-mediated transformation.

3. *FRT1*, A MUSHROOM INDUCING GENE

3.1. Isolation

FRT1 was selected from a cosmid clone bank of *Schizophyllum* genomic DNA kindly provided by Dr. R. Ullrich (University of Vermont). The vector used in construction of this clone bank contained the *TRP1* gene of *Schizophyllum* as a selectable marker; the random inserts of DNA averaged 35 kb in length. Transformation was carried out according to a protocol devised by Specht *et al.* (1988) and modified by Horton & Raper (1991a,b). Recipient cells in the transformation experiments were homokaryons, wild type for the mating-type genes but mutated for *TRP1*, hence tryptophan requiring. Transformants were identified by their ability to grow on selective medium in the absence of tryptophan. While screening the clone bank for developmental genes, we identified one cosmid that was able to induce the formation of fruiting bodies in the homokaryotic recipient. This clone was subsequently subcloned to produce a 1.4 kb active fragment and is currently being characterized at the molecular level. Results of our analysis of *FRT1* to date have been published (Horton & Raper, 1991b; Raper & Horton, 1991) and are summarized here.

3.2. Characterization

FRT1 appears to be transcriptionally regulated: by using it as a probe against polyadenylated RNA isolated from fruiting and nonfruiting cultures, we identified an 850 nucleotide sequence that is evident only during the early stages of mushroom development. Furthermore, *FRT1* is only one of at least three, possibly four, linked genes sharing strong sequence similarity within the genome from which it was isolated. Provisionally, we term these similar genes "homologues" of *FRT1*. Two of these homologues, called *FRT2* and *FRT3*, have been isolated and are being tested for biological activity in transformation experiments. *FRT2* was located about 20 kb from *FRT1* on the same cosmid clone. *FRT3* was found on another clone. Linkage of *FRT3* and the putative *FRT4* to the other two *FRT* genes was established by analysis of progeny of an outcross for segregation of these genes.

Sequence divergence for *FRT* genes between different strains was implicated by the results of DNA hybridization experiments in which it was shown that *FRT1* hybridizes only faintly to similar sequences in the genomes of strains other than the one from which it was isolated. Divergence was evidenced also by strain-specific polymorphisms with respect to location of restriction enzyme sites within the hybridizing sequences. As distinct from the homologues, we term these faintly hybridizing sequences in other strains "analogues" of *FRT1*. The analogues have not yet been isolated and analyzed at the molecular level. They could represent alternate alleles of the *FRT* genes.

3.3. Biological Activity

Both classical and molecular genetic studies of *FRT1* have established the following principles concerning its activity within living cells: 1) Cloned *FRT1* integrates stably when introduced into the genome of recipient cells via DNA-mediated transformation and appears to be trans-acting. 2) The integration of cloned *FRT1* has the effect of not only inducing *de novo* the formation of homokaryotic fruiting bodies in certain strains, but of enhancing the formation of dikaryotic fruiting bodies after the mating of homokaryons transformed for *FRT1*. 3) Endogenous *FRT1*, possibly in conjunction with its linked homologues, induces fruiting only when the mating-type genes are activated, i.e. "MAT-on", in either a dikaryon derived from a compatible mating or in an homokaryon carrying mutations of the mating-type genes which render them constitutively functional. These endogenous fruiting genes do not induce fruiting when the mating-type genes are not activated, i.e. "MAT-off", as in the wild-type homokaryon. 4) Cloned *FRT1* when integrated via transformation into MAT-off homokaryons with endogenous *FRT* alleles of a different type (non-self) does induce fruiting. 5) In contrast, cloned *FRT1* when integrated via transformation into MAT-off homokaryons with endogenous *FRT* alleles of the same type (self) does not induce fruiting. 6) The requirement of *FRT1* (and possibly its homologues) for the induction of fruiting is apparent from our observation that a spontaneous deletion of these sequences from the homokaryotic MAT-on genome is correlated with a switch in phenotype from fruiting to nonfruiting. 7) Functional equivalency of alternate alleles for the *FRT* genes has been demonstrated by showing that alternate alleles from another strain can serve in place of the deleted *FRT1* (and its homologues) in a dikaryon. A dikaryon derived by mating such a deletion mutant with a compatible wild-type homokaryon carrying *FRT* sequences of alternate types does fruit even though such a dikaryon lacks one set of the *FRT* genes: the alternate set of *FRT* genes from the wild-type mate complements the deficiency of the mutant mate with the *FRT* gene deletion.

3.4. Hypothesis Explaining Activity of *FRT1*

We have devised a testable hypothesis to reconcile these results. It incorporates the following features: 1) *FRT* genes exist in different allelic forms in different strains. 2) Each allele is closely linked to an allele-specific repressor. 3) All repressors are subject to inactivation by activity of the mating-type genes. 4) Cloned *FRT1* was fortuitously separated from its allele-specific repressor in the act of cloning. Accordingly, cloned *FRT1* is not subject to repression when integrated into a transformation recipient containing a repressor for a different allele of *FRT1* whereas it, along with endogenous *FRT1*, is subject to repression by its own specific repressor in recipients of the self type containing both the *FRT1* repressor and the *FRT1* allele. This would require that the repressor functions in a transacting manner to inactivate the heterologously integrated cloned *FRT1*.

3.5. Testing of Hypothesis

We are now attempting to identify and isolate the putative *FRT1* repressor by isolating clones that overlap the *FRT1* region in the appropriate direction. Our cloned *FRT1* lies approximately 0.7 kb from one end of the *Schizophyllum* insert in the originally isolated cosmid clone. The linked repressor may lie just beyond this region and be on an overlapping clone. A screening of the clone bank from which *FRT1* was isolated, using the methods of colony hybridization and restriction enzyme mapping, should identify such clones. These can then be tested for fruit-inducing activity in transformation experiments. A clone containing both *FRT1* and its specific repressor would be expected to be incapable of inducing fruiting in a transformation recipient carrying an alternate version of the *FRT*-repressor combination. The region containing the putative *FRT1* repressor could then be defined by testing subclones for ability to induce fruiting in transformation experiments. Once isolated, the repressor will be analyzed for structure and function. Meanwhile, the *FRT1* homologues are being analyzed and analogues (other alleles) will be isolated from other strains and ultimately characterized.

4. POSSIBLE TARGETS FOR *FRT* GENE ACTIVITY

It is likely that the *FRT* genes regulate the expression of other genes in the fruiting pathway. Such speculation is reasonable in view of the various mutations that are known to affect this pathway (Raper & Krongelb, 1958; Perkins & Raper, 1970; Springer & Wessels, 1989). Good candidates for genes regulated by *FRT* genes may be found among the *Sc* genes which are transcribed preferentially at the time of fruiting (Mulder & Wessels, 1986; Wessels, 1991). A primary target might be a homologue of the fruiting specific *Sc7* gene that we have located 2 kb away from the *FRT1* gene by DNA hybridization experiments. This possibility will be explored first by determining whether or not this *Sc7* homologue has a transcript that is temporally regulated with respect to fruiting. If so, it will be analyzed further at the molecular level.

Our primary objective is to understand the relationships among genes regulating mushroom development in *S. commune*. An understanding of these genes in this model system may be applicable to an understanding of comparable genes in edible species.

5. STRATEGY FOR ISOLATING GENES COMPARABLE TO THE *FRT* GENES OF *SCHIZOPHYLLUM* FROM OTHER, EDIBLE SPECIES OF BASIDIOMYCETES

Prospects for isolating mushroom-inducing genes from the genomes of other Basidiomycetes are dependent upon a fuller knowledge of the molecular mechanisms of *FRT* gene action in *S. commune*. We must first resolve the questions as to whether or not the *FRT1* homologues and analogues are biologically active and if so how they function in the regulation of mushroom development. We must also identify, isolate, and analyze the putative

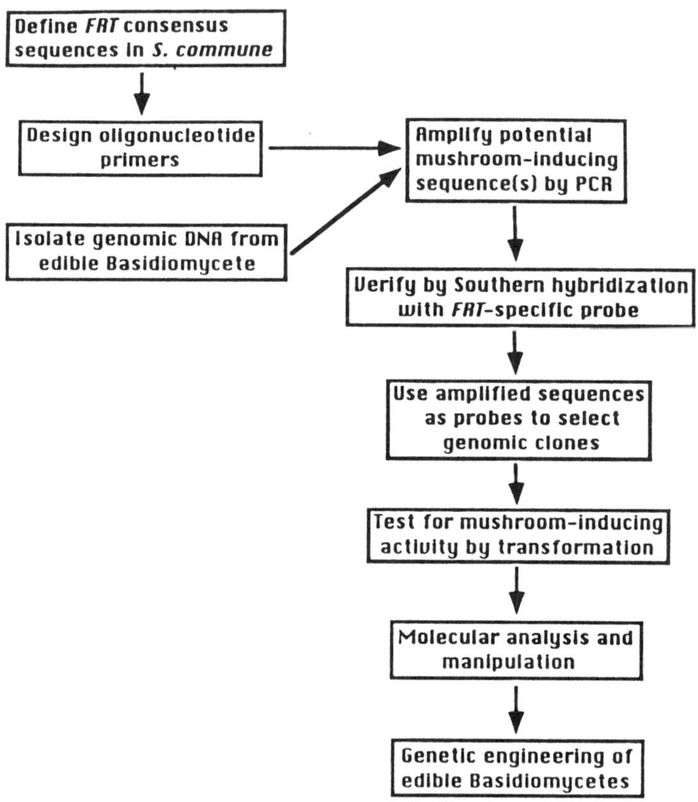

FIGURE 1. Proposed strategy for isolation of mushroom-inducing genes from edible Basidiomycetes.

repressors specific to the *FRT* alleles. In addition, it will be important to identify primary targets for *FRT* gene action and understand something about the role of the mating-type genes in determining expression of the *FRT* genes themselves. Such knowledge will be central to the design of effective strategies for identifying, isolating, and characterizing comparable genes in edible Basidiomycetes. Meanwhile, it is not unreasonable to speculate that sequences similar to *FRT1* do exist in related species.

A preliminary survey of the genomes of other species for the presence of *FRT*-like sequences could be accomplished by DNA hybridization analyses according to the methods of Southern (1975), in which the *FRT1* gene would be used as a probe against restriction enzyme digests of genomic DNAs of a variety of other Basidiomycetes. Any positive results would identify species for further investigations, but negative results would not be definitive: DNA sequences must be reasonably similar for detection by this method. A more reliable approach would employ the Polymerase Chain Reaction, known as PCR (Saiki, 1990) to amplify those parts of a gene that have been shown to be conserved and important for function. For *FRT* genes, an optimal choice of the sequence to be amplified would be based upon comparative sequence analyses and *in vitro* mutagenesis experiments for several members of the *FRT* gene family of *S. commune*. Oligonucleotide primers corresponding to important consensus sequences could then be designed for use in Polymerase Chain Reactions to amplify related sequences from the genomic DNA of other Basidiomycetes (see Fig. 1).

The amplified sequences should then be checked for correspondence to the desired sequence: the PCR procedure sometimes results in the amplification of artifactual sequences. The desired sequence could be distinguished by Southern blot analyses comparing hybridization of the amplified sequences to the *FRT*-gene probe. A PCR product would represent only a portion of the coding sequence of the putative mushroom-inducing gene and would therefore not be useful for testing *in vivo* activity via transformation. Hence, the amplified sequence would be used as a probe to screen a genomic library for clones containing the corresponding *FRT* gene. Selected clones would be tested for function in transformation experiments and subsequently subcloned, using Southern analyses to identify the region for similarity to the *FRT* gene.

If the model proposed for function of *FRT* genes in *S. commune* is correct and applicable to other species, the cloned homologues from other species would have to be separated from their allele-specific repressors and be introduced into non-self genomes in order to elicit a fruiting response in transformation. It is possible, however, that *FRT* genes operate by a different mechanism in other species.

Demonstration of a regulatory role for a *FRT* gene isolated from an edible Basidiomycete would be followed by molecular analyses to characterize and manipulate the sequence. As with the *FRT1* gene of *S. commune*, the cloned sequence linked to its own promoter might enhance the normal fruiting process in transformants for this gene and thus improve crop production. Transformation with constructs, however, in which the gene is linked to a stronger promoter could result in an overexpression of the *FRT* gene and greater mushroom yield. Constructs incorporating an inducible promotor might be useful in producing synchronous flushes of mushrooms, a definite advantage in harvesting the crop.

The application of molecular genetic techniques to other species of Basidiomycetes is critical to this strategy. The related Basidiomycete *Coprinus cinereus* could be used in initial experiments to test this approach. Although not commercially valuable as an edible species, *C. cinereus* is an excellent candidate for such tests because all the necessary techniques of molecular genetics, including DNA-mediated transformation, have been developed for this species (Mutasa *et al.*, 1990). A *FRT* gene isolated from *C. cinereus* could be tested for function *in vivo* in transformation experiments and subsequently analyzed using the techniques of *in vitro* mutagenesis.

Such techniques are not yet available in edible species. For example, transformation has not been reported for the commercial mushroom *Agaricus bisporus* despite intensive efforts by Royer & Horgen (1991) and Challen *et al.* (1991). Although the same is true for another edible mushroom, *Agrocybe aegerita*, genes that are preferentially expressed during the formation of fruiting bodies have been isolated in this species (Salvado & Labarére, 1990). These genes could prove to be similar to the Sc genes of *S. commune* which appear to be expressed in the same developmentally-regulated fashion (Mulder & Wessels, 1986). They represent potential targets for regulatory sequences such as *FRT1*.

If gene regulation of mushroom development is to be understood at the molecular level, the interaction between regulatory elements and their target genes will have to be characterized. The isolation of mushroom-inducing genes and their potential targets from *S. commune* represent important steps toward that goal. Conservation of similar developmental genes in related species is not an unreasonable concept in view of the many precedents known for conservation of genes regulating important events in the cell cycle of yeast and humans.

ACKNOWLEDGEMENTS

The work discussed in this chapter was supported by grants to C.A. Raper from the following sources: the United States National Institutes of Health (R01GM35798); the University of Vermont Committee on Research and Scholarship (BSC190-5); the American Cancer Society, Vermont Division; the United States Department of Agriculture (97-37301-6346); and the Lucille P. Markey Charitable Trust.

REFERENCES

CHALLEN, M.P., RAO, B.G. & ELLIOTT, T.J. (1991). Transformation strategies for *Agaricus*. In *Genetics and Breeding of Agaricus*, pp. 129-134. Edited by L.J.L.D. Griensven. Wageningen, The Netherlands: Pudoc.

HORTON, J.S. & RAPER, C.A. (1991a). Pulsed-field gel electrophoretic analysis of *Schizophyllum commune* chromosomal DNA. *Current Genetics* **19**, 77-80.

HORTON, J.S. & RAPER, C.A. (1991b). A mushroom inducing DNA sequence isolated from the Basidiomycete *Schizophyllum commune*. *Genetics* **129**, 707-716.

LESLIE, J.F. & LEONARD, T.J. (1979). Three independent genetic systems that control initiation of a fungal fruiting body. *Molecular and General Genetics* **171**, 257-260.

MULDER, G.H. & WESSELS, J.G.H. (1986). Molecular cloning of RNAs differentially expressed in monokaryons and dikaryons of *Schizophyllum commune* in relation to fruiting. *Experimental Mycology* **10**, 214-227.

MUTASA, E.S., TYMAN, A.M., GOTTGENS, R., MELLON, F.M., LITTLE, P.F.R. & CASSELTON, L.A. (1990). Molecular organization of an A mating-type factor of the basidiomycete fungus *Coprinus cinereus*. *Current Genetics* **18**, 223-229.

NIEDERPRUEM, D.J. & WESSELS, J.G.H. (1969). Cyto-differentiation and morphogenesis in *Schizophyllum commune*. *Bacteriological Reviews* **33**, 505-535.

NOVOTNY, C.P. & STANKIS, M.M. (1991). The *Aa* mating-type locus of *Schizophyllum commune*. In *More Genetic Manipulations in Fungi*. Edited by J. Bennett. New York: Academic Press, in press.

PERKINS, J.H. & RAPER, J.R. (1970). Morphogenesis in *Schizophyllum commune*. III. A mutation that blocks initiation of fruiting. *Molecular and General Genetics* **106**, 151-154.

RAPER, C.A. (1983). Controls for development and differentiation of the dikaryon in Basidiomycetes. In *Secondary Metabolism and Differentiation in Fungi*. Edited by J. Bennett & A. Ciegler. New York: Marcel Dekker.

RAPER, C.A. (1988). *Schizophyllum commune*, a model for genetic studies of the *Basidiomycotina*. In *Genetics of Pathogenic Fungi*, pp. 511-522. Edited by G.S. Sidham. Vol. 6, *Advances in Plant Pathology*, edited by D.S. Ingrams & P.H. Williams.

RAPER, C.A. & HORTON, J.S. (1991). A DNA sequence inducing mushroom development in *Schizophyllum*. In *Genetics and Breeding of Agaricus*, pp. 120-125. Edited by L.J.L.D. Griensven. Wageningen, The Netherlands: Pudoc.

RAPER, J.R. (1965). *Genetics of Sexuality in Higher Fungi*, 283 pp. New York: Ronald Press.

RAPER, J.R. & KRONGELB, G. (1958). Genetic and environmental aspects of fruiting in *Schizophyllum commune*. *Mycologia* **30**, 707-740.

RAUDASKOSKI, M. & VAURAS, R. (1982). A scanning electron microscope study of fruit-body differentiation in the basidiomycete *Schizophyllum commune*. *Transactions of the British Mycological Society* **78** (3), 475-481.

ROYER, J.C. & HORGEN, P.A. (1991). Towards a transformation system for *Agaricus bisporus*. In *Genetics and Breeding of Agaricus*, pp. 135-139. Edited by L.J.L.D. Griensven. Wageningen, The Netherlands: Pudoc.

SAIKI, R.K. (1990). Amplification of genomic DNA. In *PRC Protocols: A Guide to Methods and Applications*, pp. 13-20. Edited by M.A. Innis, D.H. Gelfand, J.J. Sninsky & T.J. White. San Diego: Academic Press.

SALVADO, J.C. & LABARÉRE, J. (1991). Isolation of transcripts preferentially expressed during fruit-body primordia differentiation in the

Basidiomycete *Agrocybe aegerita. Current Genetics* **20**, 205-210.

SCHWALB, M.N. (1978). Regulation of fruiting. In *Genetics and Morphogenesis in Basidiomycetes*, pp. 135-165. Edited by M.N. Schwalb & P.G. Miles. New York: Academic Press.

SOUTHERN, E.M. (1975). Detection of specific sequences among fragments separated by gel electrophoresis. *Journal of Molecular Biology*, **98**, 503-517.

SPECHT, C.A., MUNOZ-RIVAS, A., NOVOTNY, C.P. & ULLRICH, R.C. (1988). Transformation of *Schizophyllum commune*, an analysis of parameters for improving transformation frequencies. *Experimental Mycology* **12**, 357-366.

SPRINGER, J. & WESSELS, J.G.H. (1989). A frequently occurring mutation that blocks expression of fruiting genes in *Schizophyllum commune*. *Molecular and General Genetics* **219**, 486-488.

STANKIS, M.M., SPECHT, C.A. & GIASSON, L. (1990). Sexual incompatibility in *Schizophyllum commune*; from classical genetics to a molecular view. In *Seminars in Developmental Biology*, Volume 1 (3), *Developmental Systems in Fungi*, pp. 195-206. Edited by C.A. Raper & D.I. Johnson. Philadelphia: Saunders Scientific Publications.

VOLZ, P.A. & NIEDERPRUEM, D.J. (1969). Dikaryotic fruiting in *Schizophyllum commune* Fr.: Morphology of the developing basidiocarp. *Archives of Microbiology* **68**, 246-258.

WESSELS, J.G.H. (1969). Biochemistry of sexual morphogenesis in *Schizophyllum commune*: Effect of mutations affecting the incompatibility system on cell-wall metabolism. *Journal of Bacteriology* **98**, 697-704.

WESSELS, J.G.H. (1987). Mating-type genes and the control of expression of fruiting genes in basidiomycetes. *Antonie van Leeuwenhoek Journal of Microbiology* **53**, 303-317.

WESSELS, J.G.H. (1991). Hydrophobin genes in mushroom development. In *Genetics and Breeding of Agaricus*, pp. 114-119. Edited by L.J.L.D. Griensven. Wageningen, The Netherlands: Pudoc.

CHAPTER 15

EDIBLE MUSHROOMS: ATTRIBUTES AND APPLICATIONS

John A. Buswell and Shu-ting Chang

Department of Biology, The Chinese University of Hong Kong, Shatin, New Territories, Hong Kong.

1. INTRODUCTION

Throughout recorded history, mankind has savoured the desirable flavours, acknowledged the nutritional value, and recognized the medicinal and tonic properties of mushrooms. Mushrooms were regarded by the early civilizations of Egypt and Rome as a special delicacy, and were perceived by the latter as the "Food of the Gods". A clear indication of the high esteem in which mushrooms were held by different cultures is provided by the practice of restricting the use and consumption of many species to the rulers of the time. The medicinal and analeptic qualities of mushrooms, the scientific bases for which we are only just beginning to understand, have long been appreciated by the Chinese. Legend states that Chinese Emperors consumed *Lentinus edodes* in large quantities to fend off old age (Claydon, 1984). Other accounts tell us that the ancient Japanese courts valued these mushrooms for their aphrodisiac properties. "The growing sites were well hidden and heavily guarded". One species of mushroom, *Ganoderma*, is cultivated today strictly for its medicinal benefits. Furthermore, it is now realized that mushrooms have an important role to play in the recycling of organic wastes, thereby relieving environmental pollution. Since mushrooms possess the enzyme complexes which enable them to attack the cellulose, hemicellulose and lignin components of plant cell walls, they can grow on a wide variety of lignocellulosic materials. Thus, edible mushroom cultivation offers a mechanism for the effective upgrading of the huge quantities of waste residues generated annually through the activities of the agricultural, forest and food industries.

2. NUTRITIONAL VALUE OF MUSHROOMS

Most of the nutritional evaluations of edible mushrooms are based on chemical composition and do not take into account factors such as digestibility and/or nutrient availability to the human body. Nevertheless, the assessment of nutritional significance based on chemical composition may serve as a helpful reference for general comparison although quantitative evaluations of the nutritional value of edible mushrooms are subject to many variables including strain differences, the cultivation conditions, stage of cultivation and inaccuracies inherent in different methods of analysis (Crisan & Sands, 1978).

2.1. Crude Protein

Quantitative compositional analyses of several important cultivated species have revealed edible mushrooms to be a highly nutritious foodstuff that is rich in high quality protein. Crude protein estimates of mushrooms are generally calculated from the nitrogen content, as determined by the Kjeldahl method, using a conversion factor of 4.38 instead of 6.25 normally adopted for other foods (Crisan & Sands, 1978). This lower conversion factor is generally accepted as rendering a closer approximation of the crude protein content of mushrooms since studies of crude mushroom protein suggest a low coefficient of digestibility (Crisan & Sands, 1978).

From the crude protein values for several cultivated mushroom species shown in Table 1, it is evident that large variations occur even among different strains of the same species. Protein content may vary from as little as 4-8% for species of *Auricularia* to as high as 43% for *Volvariella volvacea*, based on dry weight (Crisan & Sands, 1978; Chang *et al.*, 1981; Li & Chang, 1982). Maggioni *et al.* (1968) reported that *Agaricus bisporus* produced from

TABLE 1. Crude protein, carbohydrate, fat and fibre content of cultivated mushrooms.

Mushroom	Protein (N x 4.38)	Carbohy-drate (Total)	Fat	Fibre	Energy Value (kcal)	Reference
Agaricus bisporus	23.9-34.8	51.3-62.5	1.7-8.0	8.0-10.4	328-381	Crisan & Sands (1978)
Agaricus campestris	33.2	56.9	1.9	8.1	354	Crisan & Sands (1978)
Lentinus edodes	13.4-17.5	67.5-78.0	4.9-8.0	7.3-8.0	387-392	Crisan & Sands (1978)
Pholiota nameko	20.8	66.7	4.2	6.3	372	Crisan & Sands (1978)
Pleurotus ostreatus	10.5-30.4	57.6-81.8	1.6-2.2	7.5-8.7	345-367	Crisan & Sands (1978)
Pleurotus sajor-caju	9.9-26.6	50.7-54.4	2.0-7.7	10.3-17.5	300-337	Bano & Rajarathnam (1982); Bano *et al.* (1981); El-Kattan *et al.* (1991)
Pleurotus florida	8.7-37.2	56.6-58.0	1.7-5.8	9.0-14.5	265-336	Bano & Rajarathnam (1982); Bano *et al.* (1981); El-Kattan *et al.* (1991)
Volvariella volvacea	21.3-43.0	50.9-60.0	0.7-6.4	4.4-13.4	254-374	Crisan & Sands (1978)
Volvariella diplasia	28.5	57.4	2.6	17.4	304	Crisan & Sands (1978)
Flammulina velutipes	17.6	73.1	1.9	3.7	378	Crisan & Sands (1978)
Auricularia spp.	4.2-7.7	79.9-87.6	0.8-9.7	11.9-19.8	347-384	Crisan & Sands (1978)

Values are expressed as percentage of dry weight except energy values which are presented as Kcal per 100g dry weight.

"Hauser" or "Somycel 87" spawn contained 21% and 27-35% crude protein respectively. These authors also reported that fruit bodies of the latter strain harvested in later flushes were 15-25% richer in protein. Furthermore, as shown for *V. volvacea* by Li & Chang (1982), crude protein contents of mushrooms vary widely according to the stage of fructification. Thus, the highest crude protein content (30%) was detected in the 'button' stage while levels were much lower (ca 20%) in the subsequent 'egg', 'elongation' and 'mature' stages. In the case of *Pleurotus sajor-caju*, supplementation of a rice straw substrate with cotton seed powder (Bano & Rajarathnam, 1982), or with alfalfa and soya bean meal (Zadrazil, 1980), also increased the nitrogen content of the fruit bodies.

Thus, although the protein contents of cultivated mushroom species rank below most animal meats, values compare very favourably with those of common vegetables, e.g. potato (7.6%) and cabbage (18.4%), and the staple cereals rice (7.3%) and wheat (13.2%).

2.2. Amino Acid Composition

When assessing the nutritive value of the protein, account must also be taken of the quality of the protein. In this context, the proteins of commonly cultivated mushrooms contain all the essential amino acids (Table 2) as well as most commonly occurring non-essential amino acids and amides, and are especially rich in lysine and leucine which are lacking in most staple cereal foods (Chang, 1980). However, mushroom proteins are similar to those of legumes in containing, among the essential amino acids, only low levels of tryptophan and methionine. The amino acid composition of mushrooms may alter according to the composition of the growth substrate without any apparent change in the crude protein content. Thus, mushrooms cultivated on composts supplemented with urea plus ammonium sulphate exhibited a lower total amino acid content compared to mushrooms grown on composts supplemented with inorganic nitrogen alone. Furthermore, levels of methionine, aspartic acid, valine and alanine increased while proline and arginine content decreased with urea supplementation (Maggioni *et al.*, 1968). These compositional changes became less pronounced in successive flushes. In addition to the common amino acids, various species of edible mushrooms are reported to contain less common amino acids and related nitrogenous compounds such as methionine sulphoxides, β-alanine, cystic acid, hydroxyprolines, aminoadipic acid, phosphoserine, cystathione, canavanine, creatinine, citrulline, ornithine, glucosamine and ethanolamine

TABLE 2. Essential Amino Acid Composition of Some Edible Mushrooms.

Amino Acid	*A. bisporus*	*L. edodes*	*P. ostreatus*	*V. volvacea*	*V. diplasia*
Isoleucine	200-366	218	266-267	193-261	491
Leucine	329-580	348	390-610	248-346	312
Lysine	357-527	174	250-287	427-650	384
Methionine	41-126	87	90-97	78-94	80
Phenylalanine	186-340	261	216-233	159-285	437
Threonine	243-366	261	264-290	209-307	375
Valine	112-420	261	309-326	298-414	607
Tryptophan	91-413	nd	61-87	86-112	98
Histidine	0-179	87	87-107	84-341	187
Total essential amino acids	1559-3317	1697	1933-2304	1782-2810	2971
Total amino acids	4607-7376	4962	5169-5747	4513-6332	nd

Values are expressed as mg amino acid per g corrected crude protein nitrogen. nd: not determined.
Data for *V. volvacea* from Li & Chang (1982); data for other mushrooms from Crisan & Sands (1978).

(Crisan & Sands, 1978).

2.3. Fat

The crude fat content of several edible mushroom species is shown in Table 1 and varies from less than 1.0% to almost 10.0% on a dry weight basis. This crude fat includes representatives of all classes of lipid compounds including free fatty acids, mono- di- and tri-glycerides, sterols, sterol esters and phospholipid (Crisan & Sands, 1978). Between 72% and 85% of the total fatty acids are unsaturated, the high content of which is due mainly to linoleic acid. This essential fatty acid accounts for between 54-76% of the total fatty content in *L. edodes*, and for 69% and 70% of the fatty acids in *A. bisporus* and *V. volvacea*, respectively (Table 3). The high levels of linoleic acid and the relatively low proportion of saturated fatty acids compared with animal fats is a significant contributor to the health value of mushrooms.

2.4. Carbohydrate and Fibre

Fresh mushrooms contain relatively large amounts of carbohydrate and fibre ranging from 51-88% and 4-20% respectively on a dry weight basis (Table 1). The carbohydrate component may consist of a wide variety of compounds including pentoses, methyl pentoses, hexoses, disaccharides, amino sugars, sugar alcohols and sugar acids (Crisan & Sands, 1978). Although the "mushroom sugar" trehalose is considered to be present in all mushrooms, it only occurs in significant amounts in young specimens and is hydrolyzed to glucose as the mushroom matures (Birch, 1973). Water soluble polysaccharides from the fruit bodies of mushrooms have attracted a great deal of interest recently because of their reported anti-tumour activity (see Section 4).

A major constituent of the fibre content of edible mushrooms is chitin, a polymer of N-acetylglucosamine and a structural component of the fungal cell wall. Fibre contents range from 3.7% in *F. velutipes* to between 11.9-19.8% on a dry weight basis in species of *Auricularia*. Reported fibre content of other major cultivated mushrooms varies from 7.5-17.5% in *Pleurotus* spp., 8.0-10.4% in *A. bisporus*, 7.3-8.0% in *L. edodes* and 4.4-13.4% in *V. volvacea*. This high fibre content is yet another health contributing factor of edible mushrooms. Fibre has long been recognized as an important component of a balanced and healthy diet and epidemiological data indicate that

TABLE 3. Fatty Acid Content of Some Cultivated Edible Mushrooms.

Mushroom	% Distribution of fatty acids		Fatty acid (% of total fatty acids)					
	Saturated	Unsaturated	14:0	16:0	16:1	18:0	18:1	18:2
A. bisporus	19.5 (0.60)	80.5 (2.50)	0.86	11.75	1.32	5.36	3.57	69.22
L. edodes	19.9-27.9 (0.36-0.48)	72.1-80.1 (0.94-1.68)	0.07-0.83	11.31-20.94	1.81-3.56	1.66-3.21	5.23-6.53	53.63-76.25
P. sajor-caju	20.7 (0.33)	79.3 (1.27)	0.59	16.42	1.42	3.00	12.29	62.94
V. volvacea	14.6 (0.44)	85.4 (2.56)	0.48	10.50	0.62	3.47	12.74	69.91
A. auricula	25.8 (0.34)	74.2 (0.96)	0.69	17.30	1.12	7.35	31.60	40.39
T. fuciformis	22.8 (0.14)	77.2 (0.46)	0.09	17.20	2.37	3.11	38.83	27.98

Figures in parentheses denote per cent of fatty acid per dry weight. Source: Huang *et al.* (1989).

populations on a fibre-deficient diet had a higher incidence of colonic cancer, coronary disease and other illnesses than populations eating high-fibre diets (Burkitt *et al.*, 1972). High fibre diets fed to diabetic patients reduced their daily insulin requirements and stabilized their blood glucose profile, possibly by delaying the rate of glucose absorption and/or delaying gastric emptying (Anderson & Ward, 1979).

2.5. Vitamins

Mushrooms appear to be good sources of several vitamins including thiamine, riboflavin, niacin, biotin and ascorbic acid although species differ considerably in the amount of activity they exhibit for a specific vitamin (Crisan & Sands, 1978) (Table 4). Analyses of several cultivated mushrooms have revealed that *P. nameko* contains relatively high levels of thiamine followed by *L. edodes* and *F. velutipes* (Food & Agriculture Organization, 1972). Riboflavin

TABLE 4. Vitamin content of some cultivated species of edible mushrooms.

Mushroom	Thiamine	Riboflavin	Niacin	Ascorbic acid	ProvitD2
A. bisporus	1.0-8.9	3.7-5.0	42.5-57.0	26.5-81.9	0.23
L. edodes	7.8	4.9	54.9	9.4	0.06-0.27
P. ostreatus	4.8	4.7	108.7	0	nd
V. volvacea	0.32-1.2	1.63-3.3	47.55-91.9	20.2	0.47
F. velutipes	6.1	5.2	106.5	46.3	nd
P. nameko	18.8	14.6	72.9	0	nd

Values determined using fresh material and expressed as mg/100g (dry weight) except in the case of provitamin-D2 values which are expressed as % of dry material.
nd: not determined.
Data from Crisan & Sands (1978), Li & Chang (1982) and Bano & Rajarathnam (1982).

content varied from 1.63mg/100g in *V. volvacea* to 14.6mg/100g in *P. nameko*. Recorded niacin levels range between 42.5mg/100g in *A. bisporus* to 108.7mg/100g in *P. ostreatus*. Biotin (1.7mg), pantothenic acid (22.7mg) and vitamin K have also been reported in *A. bisporus* (Altamura *et al*, 1967). Vitamin B_{12} (1.4 mg/kg dry weight) was detected in *P. ostreatus* (Shivrina *et al*, 1965). Vitamin A (retinol) activity is relatively uncommon although several mushrooms, e.g. *Tremella fusiformis*, *Auricularia polytricha*, *A. bisporus*, contain detectable amounts of provitamin A measured as β-carotene equivalent (Watt & Merrill, 1963; Food & Agriculture Organization, 1972). Similarly, vitamin D activity is rare in mushrooms (Crisan & Sands, 1978) but many contain ergosterol which can be converted to vitamin D by UV irradiation (Ramsbottom, 1953). Of several edible species examined for sterol content, *V. volvacea* had the highest provitamin-D2 content on a dry weight basis (0.47%) followed by *L. edodes* (0.27%) and *A. bisporus* (0.23%), while *T. fusiformis* had the least (0.01%) (Huang *et al.*, 1985). *P. ostreatus* (Bano & Rajarathnam, 1982), *P. sajor-caju* and *Auricularia auricula* (Chang & Miles, 1989) also contain detectable amounts. The mature stage of *V. volvacea* had a higher content of provitamin-D2 than the egg stage and a higher content was found in the cap than in the stalk, both in the mature and in the egg stages (Li & Chang, 1982).

2.6. Minerals

Mushrooms are a good source of minerals which are taken up from the substrate by the growing mycelium and translocated to the sporophores (Table 5). The major mineral elements present are potassium, which is particularly abundant, phosphorus, sodium, calcium and magnesium. Together these elements constitute between 56-70% of the total ash content (Chang & Miles, 1989). The content of phosphorus and calcium, essential for human nutrition, is often higher than in many fruits and vegetables (El-Kattan *et al.*, 1991). Minor mineral constituents include iron, copper, zinc, manganese and cobalt although in the case of iron less than one-third of the total content may be in a nutritionally-available form (Anderson & Fellers, 1942). Some species, e.g. *Pleurotus*, have a tendency to accumulate zinc in the sporophore (Bano & Rajarathnam, 1982), and undesirable metals such as cadmium and lead have also been detected in several types of cultivated mushrooms (Kikuchi *et al.*, 1984). However, levels fall well within the prescribed limits accepted by the Food and Agriculture and World Health Organizations (Bano & Rajarathnam, 1982).

2.7. Nutritional Evaluation

Quantitative data relating to the nutritive value of mushrooms is sparse. In the absence of feeding trials, alternative methods have been used to determine or predict the nutritional value of foods based on their content of essential amino acids (Crisan & Sands, 1978). The Essential Amino Acid Index (EAA Index) rates dietary protein in terms of an essential amino acid pattern based on known adult human dietary requirements. The Amino Acid Score (Chemical Score) is the amount of the most limiting amino acid in the food protein expressed as a percentage of the same amino acid present in the reference protein. In an attempt to resolve the difficulties inherent in comparisons between those mushrooms containing small amounts of high quality protein with those containing larger amounts of a protein of lesser nutritional quality, Crisan & Sands (1978) proposed the use of a Nutritional Index calculated as:

$$\text{Nutritional Index} = \frac{(\text{EAA index x percentage protein})}{100}$$

The EAA Indexes, Amino Acid Scores and Nutritional Indexes for various mushrooms and other foods are shown in Table 5. EAA Indexes and Amino Acid Scores of the most nutritive mushrooms (highest values) rank in potential nutritive value with those of meat and milk and are significantly higher than those for most legumes and vegetables. The least nutritive mushrooms rank appreciably lower but are still comparable to some common vegetables.

3. BIOCONVERSION OF LIGNOCELLULOSIC AND OTHER ORGANIC WASTES

Huge quantities of lignocellulosic and other organic waste residues are generated annually through the activities of the agricultural, forest and food processing industries. Today, there is considerable pressure to develop processes for the rational treatment and/or disposal of these residues in ways which have minimum impact on the environment. Currently, much of this

TABLE 5. Mineral content of some cultivated edible mushroom species.

Mushroom	Ca	P	K	Mg	Fe	Na
A. bisporus	23-71	790-1429	2849-4762	135	0.2-19	106-156
L. edodes	98-118	476-650	1246	nd	8.5-30	61
F. velutipes	19	278	2981	nd	11.1	278
P. ostreatus	33-79	1348	3793	140-146	15.2	837
P. sajor-caju	20-24	760-1084	3260-5265	nd	12.5-124	165-184
V. volvacea	35-347	978-1337	2005-6144	141-224	6-224	156-347
P. nameko	42	771	2083	nd	22.9	63
A. polytricha	287	trace	47.3	nd	nd	nd

Values expressed as mg/100g dry weight. nd: not determined.
Data from Crisan & Sands (1978), Li & Chang (1982) and Bano & Rajarathnam (1982).

TABLE 6. Comparison of nutritive value of mushrooms with various foods.

Essential amino acid indexes	Amino acid scores	Nutritional indexes
100 pork; chicken; beef	100 pork	59 chicken
99 milk	98 chicken; beef	43 beef
98 mushrooms (high)	91 milk	35 pork
96 *V. diplasia*	89 mushrooms(high)	31 pork
91 potatoes; kidney beans	71 *V. diplasia*	28 mushroom (high)
P. ostreatus	63 cabbage	27 *V. diplasia*
88 corn	59 potatoes	26 spinach
87 *A. bisporus*	*P. ostreatus*	25 milk
86 cucumbers	53 peanuts	22 *A. bisporus*
79 peanuts	50 corn	21 kidney beans
76 spinach; soybeans	46 kidney beans	20 peanuts
74 *L. edodes*	42 cucumbers	17 cabbage
72 mushrooms	40 *L. edodes*	15 *P. ostreatus*
69 turnips	33 turnips	14 cucumbers
53 carrots	32 mushrooms (low)	11 corn
44 tomatoes	31 carrots	10 turnips
	28 spinach	9 potatoes
	23 soybeans	8 potatoes
	18 tomatoes	6 carrots
		5 mushrooms (low)

Ranking based on essential amino acid indexes, amino acid scores and nutritional indexes as calculated against the FAO reference protein pattern. Values for mushrooms represent the mean of the three highest values (high) and the three lowest values (low).
Data from Crisan & Sands (1978) and Li & Chang (1982).

material is either burnt, shredded and/or composted for landfill or improvement of soil quality (Smith *et al.*, 1988) even though these wastes constitute a potentially valuable resource. The major components of these wastes, cellulose, hemicellulose and lignin, are relatively resistant to biological degradation. However, mushrooms possess the enzyme complexes which enable them to attack and degrade these industrial and agricultural by-products thereby resulting in a highly valued food protein suitable for direct consumption and a residue which can serve as an animal feedstock or an effective soil fertilizer and conditioner. Therefore, the cultivation of edible mushrooms represents one of the most economically viable processes for the bioconversion of agricultural and industrial wastes.

3.1. Bioconversion of Organic Wastes into Edible Protein

Although mushrooms rank below most animal meats in crude protein content, an overriding advantage of mushroom protein is that it can be produced with greater biological efficiency than proteins from animal sources. It is true that, in some highly industrialized countries, cultivation of the *Agaricus* mushroom is a highly sophisticated operation requiring a sizeable capital outlay for controlled environment facilities. However, production of other mushroom species often requires relatively little in terms of large-scale equipment, facilities, capital and land, and the mushrooms themselves often have less complicated demands in terms of processing. The straw mushroom, *V. volvacea*, is commonly grown in southeast Asian countries on small, family-type farms. Perhaps the most compelling consideration is that mushrooms can be cultivated on a wide variety of inexpensive substrates/wastes including such diverse materials as cereal straws, bagasse, banana leaves, coffee grounds, sawdust and cotton wastes from textile factories (Chang, 1991). These agricultural and industrial waste products are found in abundance in those developing regions of the world with economies which are still basically agricultural. All too often, the indigenous populations inhabiting these regions suffer from protein deficiency. Since the protein content of mushrooms is relatively high, edible mushroom production for local consumption can serve to enrich the human diet in those regions which suffer from a shortage of high quality protein. Furthermore, with the import demand elsewhere for edible mushrooms, and the growing realization that mushrooms are a potential source of high value metabolites (see Section 4), they represent a valuable cash crop. Thus, in rural areas of developing countries where there are often available large quantities of waste which is ideally suited for growing some types of edible mushrooms and where large-scale capital-intensive operations are inappropriate, properly developed and managed mushroom farms can make important contributions to the nutrition and economic welfare of the people.

3.2. Bioconversion of Waste Residues into Animal Feedstock

Edible mushroom production also represents an attractive method of improving the nutritional quality of lignocellulosic wastes for use as an animal feedstock. Agricultural and forest industry byproducts and wood which is unsuitable for pulping are carbohydrate-rich residues that represent a potential source of dietary energy for ruminants (Kamra & Zadrazil, 1988). However, the feed value is limited by the low polysaccharide degradation achieved during digestion within the rumen. This restricted digestibility is due to the presence of lignin which acts as a barrier depriving the cellulolytic and hemicellulolytic enzymes access to the polysaccharide components.

Given the annual production of the various lignocellulosic byproducts, considerable efforts are being made to develop systems for upgrading their nutritive value. Chemical and physical delignification methods have been used extensively but, along with a better understanding of the microbial physiology and biochemistry of lignin biodegradation (Buswell & Odier, 1987; Buswell, 1991), more attention is being focused on delignification treatments based on lignin-degrading fungi, including several edible species (Table 7) (Kirk, 1983). Relatively higher lignin degradation rates and consequent increases in digestibility are obtainable using cereal straws as a substrate as compared with wood, and several white-rot fungi exhibit a high capacity to increase the *in vitro* digestibility of wheat straw (Table 8) (Kamra & Zadrazil, 1988). Even so, biological delignification of wood preparations may also offer possibilities for the production of ruminant feedstuff. In southern Chile, fungal delignification of wood has been observed under natural conditions. The product of delignification, known as 'palo podrido' is a white decomposed wood which is used as an animal feed (Kamra & Zadrazil, 1988). *In vitro* digestibility of the wood is increased from 3% to 77% in some cases although the process is long and slow. More effective treatments for enhancing the digestibility of wood, straws and other lignocellulosic byproducts using ligninolytic fungi are dependent on further co-ordinated research aimed at optimization of the solid state fermentation processes involved.

3.3. Spent Mushroom Compost as a Soil Fertilizer and Conditioner

Spent mushroom compost consisting of degraded cellulose, hemicellulose and lignin serves as an effective soil fertilizer and conditioner. In addition to providing a balanced nitrogen and carbon source for plant growth, the spent compost undergoes further transformation in the soil to form humus. This material plays a central role in maintaining soil structure, and in improving soil aeration and water-holding capacity. When spent cotton-waste compost following cultivation of *V. volvacea* was used to grow lettuce, Chinese radish and tomato, vegetable yields were 3-, 3- and 7-fold higher, respectively compared to those obtained using regular garden soil (Chang & Yau, 1981).

TABLE 7. Effect of edible mushrooms on *in vitro* digestibility of lignocelluloses.

Mushroom	Substrate	Time (days)	Total weight loss (%)	*In vitro* digestibility (%)[a]	
				Before decay	After decay[b]
Pleurotus sp. Florida	Beech wood	60	17	6	35
Pleurotus sp. Florida	Reed straw	60	30	30	45
Pleurotus sp. Florida.	Sunflower stalks	60	27	41	62
L. edodes (Berk.) Sing.	Wheat straw	60	13	40	77
L. edodes (Berk.) Sing.	Birch wood	69	25	20	60

a Rumen fluid method
b Values expressed as % of decayed sample
Data from Kirk (1983)

TABLE 8. Cultivated fungi with a high capacity to increase *in vitro* digestibility of wheat straw.

Fungus	Temp. (°C)	Organic Matter (OM) loss (%)	Lignin loss (% OM)	Change in digestibility	Process efficiency
L. edodes	25	17.4	3.0	+ 24.6	+ 1.41
P. ostreatus	22	15.4	7.0	+ 22.4	+ 1.45
P. sajor-caju	30	18.7	2.7	+ 22.0	+ 1.18
Ganoderma lucidum	25	25.2	10.8	+ 25.7	+ 1.02

Process efficiency = Change in *in vitro* digestibility divided by dry matter loss during fermentation.
Data from Kamra & Zadrazil (1988).

4. MEDICINAL AND TONIC QUALITIES OF MUSHROOMS

Although mushrooms have traditionally been used in China and Japan for their medicinal and tonic properties, this aspect of mushrooms remains largely unexploited. However, there has been a recent upsurge in interest in traditional remedies for the treatment of various physiological disorders and numerous biologically active compounds have been reported in mushrooms as the result. Lists of pharmaceutical products developed from mushrooms in Japan (Table 9) and their active components (Table 10) have been compiled recently by Pai *et al.* (1990). Cosmetic products and tonic beverages have also been produced in China from *Ganoderma* mushrooms. This section describing some of the medical properties of mushrooms will emphasize the beneficial effects and will not address toxic parameters except in the context of toxicity to a biological agent that is harmful to man.

4.1. Anti-tumour Effects

Several edible fungi have been reported to exhibit anti-tumour activity

TABLE 9. Pharmaceuticals developed from mushrooms in Japan.

Name	Krestin	Lentinan	Schizophyllan
Abbreviation	PSK/PSP		
Date for sale	May 1977	December 1985	April 1986
Mushrooms species	*Coriolus versicolor* (mycelium)	*Lentinus edodes* (fruiting body)	*Schizophyllum commune*
Polysaccharide	β-1,6 branch; β-1,3; Beta-1,4 mainchain	β-1,6 branch; β-1,3 mainchain	β-1,6 branch; β-1,3 mainchain
Molecular weight	ca. 100,000	ca. 500,000	ca. 450,000
$[\alpha]_D$		+14~22°C (NaOH)	+18~24°C (H_2O)
Products	1g/package	1mg/vial	1g/2ml bottle
Administration	Oral	Injection	Injection
Indication	Cancer of digestive system, breast cancer, pulmonary cancer	Gastric cancer	Cervical cancer
1985 sale value	556 M$	85 M$	128 M$

Source: Pai *et al.* 1990

including *L. edodes, F. velutipes, P. ostreatus, A. bisporus, P. nameko, Tricholoma matsutake* and *A. auricula* (Ikekawa *et al.*, 1969; Vogel *et al.*, 1975). *V. volvacea* and *F. velutipes* contain cardiotoxic proteins, volvatoxin A and flammutoxin, which inhibit the respiration of Ehrlich ascites tumour cells (Lin *et al.*, 1974). Considerable attention has focused on lentinan, a polysaccharide extracted from *L. edodes*, the anti-cancer activity of which is reported to extend to cancer of the bowel, pancreas, gastrointestinal tract, liver, lung and the ovaries (Flynn, 1991). In a four year randomized control trial of lentinan for advanced or recurrent stomach cancer in combination with chemotherapy, the two year survival rate in the control group was 0.1%

Table 10. Pharmaceutical components of mushroom species.

	Pharmacodynamic	Component	Species
1.	Antibacterial effect	Hirsutic acid	Many species
2.	Antibiotic	E-β-methoxyacrylate	*Oudemansiella radicata*
3.	Antiviral effect	Polysaccharide, Protein	*Lentinus edodes* and *Polyporaceae*
4.	Cardiac tonic	Volvatoxin, Flammutoxin	*Volvariella*
5.	Decrease cholesterol	Eritadenine	*Collybia velutipes*
6.	Decrease level of blood	Peptide glycogen, Ganoderan, glucan	*Ganoderma lucidum*
7.	Decrease blood pressure	Triterpene	*Ganoderma lucidum*
8.	Antithrombus	5'-AMP, 5'-GMP	*Psalliota hortensis*
9.	Inhibition of PHA	r-GHP	*Psalliota hortensis, Lentinus edodes*
10.	Antitumor	β-glucan RNA complex	Many species, *Hypsizygus marmoreus (Lyophyllum shimeji)*
11.	Increase secretion of bile	Armillarisia A	*Armillariella tabescens*
12.	Analgesic, Sedative effect	Marasmic acid	*Marasmius androsaceus*

Source: Pai *et al.* 1990

whereas in the lentinan-treated group, the two- and four-year survival rates were 9.5% and 3.8%, respectively (Taguchi *et al.*, 1985). Lentinan stimulates T-lymphocyte production (Chihara *et al.*, 1987) which is suppressed in cancer states. Furthermore, the anti-tumour action of lentinan occurred even when the agent was administered orally to mice as opposed to previous intravenous or intraperitoneal injection. *L. edodes* may also exert an anti-cancer effect by preventing *in vivo* formation of carcinogens. When dried shiitake is boiled, thiazolidine-4-carboxylic acid (TCA) is formed (Kurashima *et al.*, 1990). TCA is an effective nitrite trapping agent and may block the formation of carcinogenic N-nitroso compounds.

4.2. Anti-viral Effects

Cochran *et al.* (1967) reported that a polysaccharide fraction from *L. edodes* was active in reducing the number of lung lesions in mice caused by influenza A/SW15 virus. Other antiviral activity in *L. edodes* was mediated by the induction of interferon in the host (Takehara *et al.*, 1979). This induction of interferon was subsequently attributed to mycoviral double-stranded RNA extracted from fungal spores (Suzuki *et al.*, 1976). Another substance called KS-2 extracted from mushrooms by Fujii *et al.* (1978) was shown to have anti-viral activity by inducing interferon production. Japanese researchers recently reported that lentinan in combination with the drug AZT was more effective than AZT itself in suppressing the proliferation of the AIDS virus (Tochikura *et al.*, 1987). Pretreatment of the AIDS virus with an extract of *L. edodes* blocked infection of the target cells.

4.3. Hypocholesterolaemic Effects

Kaneda and coworkers first reported that a diet supplemented with 5% ground dry sporophores of *L. edodes* lowered average plasma cholesterol levels about 24% when fed to rats for ten weeks. The hypocholesterolaemic effect was most evident with the Donko variety of *L. edodes* which produced a 45% reduction in total plasma cholesterol. *A. bisporus,* the Koshin variety of *L. edodes, A. polytricha* and *F. velutipes* were less active (Kaneda & Tokuda, 1966). Hypocholesterolaemia was attributed to acceleration of cholesterol metabolism and increased cholesterol excretion (Tokuda & Kaneda, 1976). Chronic ingestion of *L. edodes* was also reported to reduce serum cholesterol levels in man (Suzuki & Oshima, 1976). An active hypolipidaemic principle in *L. edodes* has been identified as eritadenine which gives rise to a general response, affecting cholesterol, triglyceride and phospholipid levels (Tokuda & Kanana, 1978). More recently, Bobek *et al.* (1990) reported a hypocholesterolaemic effect of oyster mushroom (*P. ostreatus*) in rats with hereditary increased sensitivity to dietary cholesterol. The observed falls in serum cholesterol levels were due exclusively to a fall in cholesterol lipoproteins of very low density and of low density. The mushroom significantly increased the activity of lecithin cholesterol acyltransferase and decreased the activity of lipoprotein lipase in abdominal adipose tissue.

4.4. Antibiotic Effects

Compared with lower fungi, antibacterial activity among edible species is poorly documented. Benedict and Brady (1972) have listed various antibacterial antibiotics from higher fungi including polyacetylenes, phenolic compounds, purines and pyrimidines, quinones and terpenoids. Edible fungi occurred in, but did not dominate, the list of sources. Vogel *et al.* (1974) described phenolic and quinoid derivatives with antibacterial activity in *A. bisporus*. Examples of antifungal activity among edible fungi include *L. edodes* (Herrman, 1962), *Coprinus comatus* (Bohus *et al.*, 1961) and *Oudemansiella mucida* (Musilek *et al.*, 1969).

4.5. Other Biological Activities of Edible Mushrooms

In a study of impotent patients in France, most were found to suffer from erection impairment related to arteriosclerotic changes in the arteries of the penis (Virag *et al.*, 1985). Lentinan is also reported to induce the formation of prostaglandins (Flynn, 1991) some of which, e.g. prostaglandin E1, are smooth muscle relaxants which cause dilatation of the blood vessels, an important factor in erections. Thus, this effect coupled with the hypocholesterolaemic activity of lentinan may account in part for the perceived aphrodisiac properties of *L. edodes*. An alternative or supplementary explanation may involve the reported ability of the shiitake mushroom to accumulate zinc in the sporophore (Timmer *et al.*, 1990), and the recorded links between zinc and plasma testosterone levels (Antoniou *et al.*, 1977).

An inhibitor of platelet aggregation, subsequently identified as adenosine, was isolated from *A. polytricha.* Regular consumption of the fungus was thought to be responsible for the low incidence rate of atherosclerosis among Asians who consumed this fungus regularly (Markhija & Bailley, 1981).

In addition to lentinan, other hypotensive agents have been reported in edible fungi including a triterpene component of *Ganoderma lucidum* (Pai *et al.*, 1990). Extracts of *P. sajor-caju* were associated with a hypotensive action in rats (Tam *et al.*, 1986).

4.6. Tonic Qualities of Edible Mushrooms

Mycelium of edible mushrooms grown in liquid culture has been used in the making of soups and teas, a practice especially popular with edible fungi that are thought to have medicinal or tonic qualities. In China, the mycelium of the golden ear mushroom, *Tremella aurantia*, that is harvested from liquid culture is added to walnut cakes, biscuits and bread. It is also put into drinks for older people and children and added to flour for making noodles. Preliminary studies have been undertaken on the formation of an acceptable food product (tentatively called 'mycomeat') prepared by growing edible fungi on waste soybean residues from tofu production. The product formed has good taste and texture, with the flavour determined by the edible mushroom used (Chang & Miles, 1989).

5. MISCELLANEOUS APPLICATIONS

5.1. Edible Mushrooms as a Source of Flavourants

Largely as a result of the food industry's continuing demand for natural ingredients and flavours, mushrooms are receiving increasing attention recently as a source of flavour compounds (Hadar & Dosoretz, 1991). The flavour-based gastronomic appeal is, after all, one of the main reasons for consuming wild and commercially grown edible mushrooms. The profile of flavour-imparting compounds can vary markedly among species and even among varieties, and can also be influenced by culture conditions (Hadar & Dosoretz, 1991). How the various components combine to give the characteristic flavour of mushrooms is not yet clear. Certain non-volatile substances may contribute to the characteristic flavour, including L-glutamic acid, unusual amino acids, short chain fatty acids, carbohydrates, proteins and non-protein nitrogenous substances such as nucleotides (mainly guanosine monophosphate) (Hadar & Dosoretz, 1991). The chemical composition of the volatile fraction is also believed to be largely responsible for mushroom flavour. About 150 different volatile compounds have been identified in various mushroom species, representing a wide variety of chemical structures including simple aliphatic alcohols, aldehydes, ketones, esters, lactones, mono- and sesquiterpenes, and aromatics such as cinnamyl derivatives. Typical flavour compounds synthesized by mushrooms include volatiles derived from the metabolism of fatty acids, especially the 'mushroom alcohol' 1-octen-3-ol. *L. edodes* produces the cyclic sulphur-containing compound lenthionine as the primary aroma component (Chen *et al.*, 1986). Various compounds, including 3-methylbutanal, butanol, 3-methylbutanol, pentanol, hexanol, furfural, phenylacetaldehyde and α-terpineol, have been identified as minor volatile components of mushrooms (Gallois *et al.*, 1990). Submerged fermentation techniques for the production of fungal biomass that is rich in flavour are now being developed (Hadar & Dosoretz, 1991).

5.2. Mushrooms in Floristry

Mushrooms are also cultivated commercially for use in floristry (Poppe & Heungens, 1991). Methods have been developed for the large-scale production of fruit bodies of *Pleurotus, Lentinus, Pholiota, Coriolus* and *Polyporus* on wood stems or artificial substrates for ornamental purposes. Gardens adorned with gypsum toadstools may well become a thing of the past.

6. CONCLUSION

As a result of technical advances achieved during recent years, the commercial cultivation of edible mushrooms has spread to many countries throughout the world. Since cultivated mushrooms can be grown on agricultural and industrial wastes, they provide a solution to many problems of global importance including protein shortages, resource recovery and re-use, and environmental management. Edible mushrooms are a source of high quality protein which can be produced with greater biological efficiency than animal protein and therefore have important potential in less developed countries for enriching the diet of populations suffering from protein deficiency. As well as being consumed as a source of vegetable protein, many edible mushrooms are receiving additional recognition for their medicinal and tonic qualities. Significant pharmacological properties have been demonstrated including immunopotentiation, anti-tumour and hypocholesterolaemic effects. A variety of proprietary products derived from edible mushrooms, including health drinks, foods and flavourants, and even cosmetics, are already available and the market for such materials is expected to increase. Edible mushroom production also represents an attractive method of improving the nutritional quality of lignocellulosic wastes for use as an animal feedstock. Spent compost, the substrate residue left after mushroom harvesting, can be converted into feedstock for ruminants and/or used as a soil fertilizer and conditioner.

REFERENCES

ALTAMURA, M.R., ROBBINS, F.M., ANDREOTTI, R.E., LONG, L., Jr. & HASSELSTROM, T. (1967). Mushroom ninhydrin-positive compounds. Amino acids, related compounds, and other nitrogenous substances found in cultivated mushroom, *Agaricus campestris*. *Journal of Agricultural and Food Chemistry* **15**, 1040-1043.

ANDERSON, E.E. & FELLERS, C.R. (1942). The food value of mushroom *Agaricus campestris*. *Proceedings of the American Horticultural Society* **41**, 310-304.

ANDERSON, J.W. & WARD, K. (1979). High-carbohydrate high-fiber diets for insulin-treated men with diabetes mellitus. *American Journal of Clinical Nutrition* **32**, 2312-2321.

ANTONIOU, L.D., SHALHOUB, R.J., SUDHAKAR, T. & SMITH, J.C. (1977). Reversal of uraemic impotence by zinc. *Lancet* **2**, 895-898.

BANO, Z. & RAJARATHNAM, S. (1982). *Pleurotus* mushroom as a nutritious food. In *Tropical Mushrooms - Biological Nature and Cultivation Methods*, pp.363-380. Edited by S.T. Chang & T.H. Quimio. Hong Kong: Chinese University Press.

BANO, Z., BHAGYA, S. & SRINIVASAN, K.S. (1981). Essential amino acid composition and proximate analysis of the mushrooms *Pleurotus eous* and *P. florida*. *Mushroom Newsletter for the Tropics* **1**, 6-10.

BENEDICT, R.G. & BRADY, L.R. (1972). Antimicrobial activity of mushroom metabolites. *Journal of Pharmaceutical Sciences* **61**, 1820-1822.

BIRCH, G.G. (1973). Mushroom sugar. *Plant Foods Manual* **1**, 49-55.

BOBEK, P., GINTER., OZDIN,L. & CERVEN, J. (1990). Hypocholesterolemic effect of oyster mushroom (*Pleurotus ostreatus*) in rat with hereditary increased sensitivity to dietary cholesterol. *Biologia (Bratislava)* **45**, 961-966.

BOHUS, G., GLAZ, E. & SCHREIBER, E. (1961). The antibiotic action of higher fungi on resistant bacteria and fungi. *Acta Biologia Academia Scientiarum Hungarica* **12**, 1-12.

BURKITT, D.P., WALKER, A.R.P. & PAINTER, N.S. (1972). Effect of dietary fibre on stools and transit-times and its role in the causation of disease. *Lancet* **2**, 1408-1412.

BUSWELL, J.A. (1991). Fungal degradation of lignin. In *Handbook of Applied Mycology, Volume 1*, pp. 425-480. Edited by A.K. Arora, B. Rai, K.G. Mukerji & G. Knudsen. New York: Marcel Dekker, Inc.

BUSWELL, J.A. & ODIER, E. (1987). Lignin biodegradation. *CRC Critical Reviews in Biotechnology* **6**, 1-60.

CHANG, S.T. (1980). Mushrooms as human food. *Bioscience* **30**, 399-401.

CHANG, S.T. (1991). In *Handbook in Applied Mycology, Volume 3*, pp.221-240. Edited by A.K. Arora, K.G. Mukerji & M.H. Marth. New York: Marcel Dekker, Inc.

CHANG, S.T. & MILES, P.G. (1989). *Edible Mushrooms and their Cultivation*. Boca Raton: CRC Press.

CHANG, S.T. & YAU, C.K. (1981). Production of mushroom food and crop fertilizer from organic wastes. In *GIAM VI Global Impacts of Applied Microbiology*, pp.647-652. Edited by S.O. Emejuaiwe, O. Ogunbi & S.O. Sanni. New York and London: Academic Press.

CHANG, S.T., LAU, O.W. & CHO, K.Y. (1981). The cultivation and nutritional value of *Pleurotus sajor-caju*. *European Journal of Applied Microbiology and Biotechnology* **12**, 58-62.

CHEN, C.C., LIU, S.E., WU, C.M. & HO, C.T. (1986). Enzymic formation of volatile compounds in shiitake mushroom (*Lentinus edodes* Sing.). *American Chemical Society Symposium Series* **317**, 176-183.

CHIHARA, G., HAMURO, J. & MAEDA, Y.Y. (1987). Anti-tumour and metastasis: inhibitory actions of lentinan as an immunomodulator. *Cancer Detection and Prevention Supplement*, **1**, 423-443.

CLAYDON, N. (1984). Secondary metabolic products of selected Agarics. In *Developmental Biology of Higher Fungi,* pp. 561-580. Edited by D. Moore, L.A. Casselton, D.A. Wood & J.C. Franklin. Cambridge: Cambridge University Press.

COCHRAN, K.W., BENEKE, E.S. & NISHIKAWA, T. (1967). Botanical sources of influenza virus inhibitors. In *Proceedings of the Sixth Interscience Conference on Antimicrobial Agents and Chemotherapy,* pp.515-520. Washington, D.C: American Society for Microbiology.

CRISAN, E.V. & SANDS, A. (1978). Nutritional value. In *The Biology and Cultivation of Edible Mushrooms,* pp.137-165. Edited by S.T. Chang & W.A. Hayes. New York: Academic Press.

EL-KATTAN, M.H., HELMY, Z.A., EL-LEITHY, M.A.E. & ABDELKAWI, K.A. (1991). Studies on cultivation techniques and chemical composition of oyster mushrooms. *Mushroom Journal for the Tropics* **11**, 59-66.

FLYNN, V.T. (1991). Is the shiitake mushroom an aphrodisiac and a cause of longevity? In *Science and Cultivation of Edible Fungi,* pp.345-361. Edited by Maher, M.J. Rotterdam: Balkema.

FOOD AND AGRICULTURAL ORGANIZATION. (1972). Food composition table for use in East Asia. *Food Policy and Nutrition Division., Food and Agriculture Organization.* Rome: United Nations.

FUJII, T., MAEDA, H., SUZUKI,F. & ISHIDA, N. (1978). Isolation and characterization of a new antitumor polysaccharide, KS-2, extracted from culture media of *Lentinus edodes. Journal of Antibiotics (Tokyo)* **31**, 1079-1090.

GALLOIS,A., CROSS,B., LANGLOIS,D., SPINNLER, H.E. & BRUNERIE, P. (1990). Influence of culture conditions on production of flavour compounds by 29 ligninolytic Basidiomycetes. *Mycological Research* **94**, 494-504.

HADAR, Y. & DOSORETZ, C.G. (1991). Mushroom mycelium as a potential source of food flavour. *Trends in Food Science and Technology* **3**, 214-218.

HERRMAN, H. (1962). Cortinellin, eine antibiotisch wirksam Substanz aus *Cortinellus shiitake. Naturwissenschaften* **49**, 542.

HUANG, B.H., YUNG, K.H. & CHANG, S.T. (1985). The sterol composition of *Volvariella volvacea* and other edible mushrooms. *Mycologia* **77**, 959-963.

HUANG, B.H., YUNG, K.H. & CHANG, S.T. (1989). Fatty acid composition of *Volvariella volvacea* and other edible mushrooms. *Mushroom Science* **12**, 533-540.

IKEKAWA, T., UEHARA, N., MAEDA, Y., NAKAMISHI, M. & FUKUOKA, F. (1969). Anti-tumour activity of aqueous extracts of some edible mushrooms. *Cancer Research* **92**, 734-735.

KAMRA, D.N. & ZADRAZIL, F. (1988). Microbiological improvement of lignocellulosics in animal feed production-a review. In *Treatment of Lignocellulosics with White-rot Fungi*, pp.53-63. Edited by F. Zadrazil & P. Reiniger. London & New York: Elsevier.

KANEDA, T. & TOKUDA, S. (1966). Effect of various mushroom preparations on cholesterol levels in rats. *Journal of Nutrition* **90**, 371-376.

KIKUCHI, M., TAMAKAWA, K. & HIROSHIMA, Y. (1984). Survey on contents of metals in edible mushrooms. *Journal of the Food Hygiene Society of Japan* **25**, 534-542.

KIRK, T.K. (1983). Degradation and conversion of lignocelluloses. In *The Filamentous Fungi, Volume 4*. pp.266-295. Edited by J.E. Smith, D.R. Berry & B. Kristiansen. London: Edward Arnold.

KURASHIMA, Y., TSUDA, M. & SUGIMURA, T. (1990). Marked formation of thiazolidine-4-carboxylic acid, an effective nitrite trapping agent *in vivo*, on boiling of dried shiitake mushroom (*L. edodes*). *Journal of Agricultural and Food Chemistry* **38**, 1945-1949.

LI, G,S,F. & CHANG, S.T. (1982). Nutritive value of *Volvariella volvacea*. In *Tropical Mushrooms - Biological Nature and Cultivation Methods*, pp.199-219. Edited by S.T. Chang & T.H. Quimio. Hong Kong: Chinese University Press.

LIN, J.Y., LIN, Y.J., CHEN, C.C., WU, H.L., SHI, G.Y. & JENG, T.W. (1974). Cardiotoxic protein from edible mushrooms. *Nature* **252**, 235-237.

MAGGIONI, A., PASSERA, C., RENOSTO, F. & BENETTI, E. (1968). Composition of cultivated mushrooms (*Agaricus bisporus*) during the growing cycle as affected by the nitrogen source introduced during composting. *Journal of Agricultural and Food Chemistry* **16**, 517-519.

MARKHIJA, A.N. & BAILLEY, J.M. (1981). Identification of the antiplatelet substance in Chinese black tree fungus. *New England Journal of Medicine* **304**, 175.

MUSILEK, V., CERNA, V.J., SASEK, V., SEMERDZIEVA, M. &

VONDRACEK, M. (1969). Antifungal antibiotic of the basidiomycete *Oudemansiella mucida*. I. Isolation and cultivation of a producing strain. *Folia Microbiogia* (Prague) **14**, 377-387.

PAI, S.H., JONG, S.C. & LOW, D.W. (1990). Usages of mushroom. *Bioindustry* **1**, 126-131.

POPPE, J. & HEUNGENS, K. (1991). First commercial growing of ornamental mushrooms and its use in floristry. In *Science and Cultivation of Edible Fungi*, pp. 821-827. Edited by Maher, M.J. Rotterdam: Balkema.

RAMSBOTTOM, J. (1953). Mushrooms and toadstools. *Proceedings in Nutritional Science* **12**, 39-44.

SHIVRINA, A.N., KORYOKINA, L.N. & YAKIMOV, P.A. (1965). Vitamin B_{12} content of polypore and agaric fungi. In *Kormovye Belki i fiziologicheski aktivnye veshchestva dlya zhivotnovodstva,* pp.88-91. Moscow-Leningrad: Akademiya Nauk, S.S.S.R.

SMITH, J.F., FERMOR, T.R. & ZADRAZIL, F. (1988). Pretreatment of lignocellulosics for edible fungi. In *Treatment of Lignocellulosics with White-rot Fungi*, pp. 3-13. Edited by Zadrazil, F. & Reiniger, P. London & New York: Elsevier.

SUZUKI, S. & OSHIMA, S. (1976). Influence of Shii-ta-ke (*Lentinus edodes*) on human serum cholesterol. *Mushroom Science* **9**, 463-467.

SUZUKI, F., KOIDE, T., TSUNODA, A. & ISHIDA, N. (1976). Mushroom extract as an interferon inducer. I. Biological and physiochemical properties of spore extracts of *Lentinus edodes. Mushroom Science* **9**, 509-520.

TAGUCHI, T., FURUE, H. & KIMURA, T. (1985). End point results of phase III study of lentinan. *Japanese Journal of Cancer Chemotherapy* **12**, 366-378.

TAKEHARA, M., KUIDA, K. & MORI, K. (1979). Antiviral activity of virus-like particles from *Lentinus edodes* (Shiitake). *Archives of Virology* **59**, 269-274.

TAM, S.C., YIP., FUNG, K.P. & CHANG, S.T. (1986). Hypotensive and renal effects of an extract of the edible mushroom *Pleurotus sajor-caju. Life Science* **38**, 1155-1161.

TIMMER, J., PERSHERN, A. & ONDRUS, M. (1990) A nutritional analysis and development of promotional materials for shiitake producers in Wisconsin. *Shiitake News* **7**, 6-11.

TOCHIKURA, T.S., NAKASHIMA, H., KANEKO, Y., KOBAYASHI, N. & YAMAMOTO, N. (1987). Suppression of human immuno-deficiency virus replication by 3'-azido-3'-deoxythymidine in various human haematopoietic cell lines: Augmentation of the effect by lentinan. *Japanese Journal of Cancer Research (Gann)* **78**, 583-589.

TOKUDA, S. & KANANA, T. (1978). Effect of shiitake mushrooms on plasma cholesterol level in rats. *Mushroom Science* **10**, 793-796.

TOKUDA, S. & KANEDA, T. (1976). Reducing mechanism of plasma cholesterol by Shii-ta-ke. *Mushroom Science* **9**, 445-462.

VIRAG, R., BOUILLY, P. & FRYDMAN, D. (1985). Is impotence an arterial disorder? - A study of arterial risk factors in 440 impotent men. *Lancet* **1**, 181-184.

VOGEL, F.S., McGARRY, S.J., KEMPER, L.A.K. & GRAHAM, D.G. (1974). Bacteriocidal properties of a class of quinoid compounds related to the sporulation of the mushroom, *Agaricus bisporus*. *American Journal of Pathology* **76**, 165-174.

VOGEL, F.S., KEMPER, L.A., McGARRY, S.J. & GRAHAM, D.G. (1975). Cytostatic, cytocidal and potential antitumor properties of a class of quinoid compounds, initiators of the dormant state in the spores of *Agaricus bisporus*. *American Journal of Pathology* **78**, 33-48.

WATT, B.K. & MERRILL, A.L. (1963). Composition of foods. *United States Department of Agriculture., Agricultural Handbook* **8**.

ZADRAZIL, F. (1980). Influence of ammonium nitrate and organic supplements on the yield of *Pleurotus sajor-caju* (Fr.) Sing. *European Journal of Applied Microbiology* **9**, 31-35.

Index

Index

T

U

V

W

Y